"I hear and I forget. I see and I remember. I do and I understand."

— Confucius

SMART OUTBOUND

Let's do this.

~. _ .~

SMART OUTBOUND

The Outbound Sales Journal

Matt Wanty

Priority House

Published by Priority House

First published in the United States of America by Priority House
This paperback edition published in 2020.

ISBN-13: 978-0578780597
(Priority House)

ISBN-10: 0578780593

Printed by Amazon

Long live, Outbound Sales!

~. _ ,~

Dedicated to Bianca Wanty

I'm dedicating this book to my wife. Every word I've ever written she's been right there ready to read it. Every crazy idea I've ever had she's been right there willing to hear it.

My wife Bianca is the best friend I've ever had. She and I are traveling through the good and bad of life together as one, stronger, smarter and braver because we have each other (and our awesome daughter Naida).

Her love and support have been steadfast as we try to make the world a little better. I am lucky to have found such a beautiful person to spend my life with.

Without her you surely wouldn't be reading this book.

I love you, bunny!

Contents

1

Introduction

Hi, my name is Matt Wanty. I'm an outbound sales strategist and 3x Founder. You may have seen some of my LinkedIn content that has reached millions of sales people across the globe. Here's what you should know about this book. Over the last two years I've spoken with almost one thousand sales people from all over the world in almost every industry. In these conversations we would share best practices and tactical tips. During the same time, I've worked individually with over one hundred sales people helping them to refine their sales messaging. As you can probably imagine, I've learned a lot more about outbound sales along the way. The total body of interactions became a series of data points for me, about what's working and what's not in outbound sales. This experience has led me to writing my second book to help sales people crack more accounts and crush quota. It's filled with everything I know about succeeding with outbound sales in 2021.

If I was a dramatic guy, I'd tell you outbound sales is a whole new ball game. I'm not that dramatic. But I will say it's certainly a much different time than it was just 5 or 10 years ago. Why? Two simple reasons: social media and video. These new mediums have multiplied the number of ways sales people can get in front of decision makers. They've also enhanced it. Don't get me wrong, the old mediums like cold calling, cold emails, direct mail and even voicemails are still viable. Of course, it all depends on what you're selling but many sales teams are still succeeding with these channels.

In outbound sales it doesn't really matter how good the medium if you don't have the skills to pull it off. Some people excel with cold calling. Some reps have the patience and tact to succeed with social media. Others are really good at writing email copy. Some people shine with video. In the end you have to find a mix that works best for you and I'd be lying if I told you it was easy. Determination and a willingness to be uncomfortable are required if you want to develop strong outbound sales skills.

The harsh reality with outbound sales is that it's really hard to succeed. Because of a multitude of circumstances, some people new to outbound have an impossible hill to climb. Lack of training, ineffective sales messaging, bad outbound channel strategy, uncompetitive products or services are just a few of the things that can make it difficult for reps to be successful. I hope this book helps anyone who is in a tough situation excel with outbound sales.

The days of just pounding the phone are over. For the most part that's a bad B2B strategy because you'll only reach the small portion of decision makers who pick up cold calls. Other decision makers may read emails, some may be more willing to interact on social media and others might even watch your video. Today the best outbound sales strategies are Omni-channel. Please don't misunderstand me, if you find an outbound tactic or medium that's working at a high rate. Hammer it. Nothing lasts forever and you need to seize the opportunity so you can get the most out of it for you and your company.

I don't know about you but selling has always been something I could find myself doing naturally. Once I get rolling I usually don't have a problem saying words that get decision makers interested. When you think about life, we all sell ourselves every day. We sell to our family. We sell to our friends. Whether we're trying to sell smiles or missing report card stories, at a very early age we're taught how to get people on our side. Heck, my three year-old is a master of it and she can barely talk. There's a fine line when it comes to manipulation in sales. But selling itself is manipulation. We're literally trying to get someone to do something they may not want to do at first. Matter of fact, it's our job to make it happen. What makes all the difference in the world is the way you go about it. I hope this book helps you find ways to create sales opportunities without having to be a jerk about it.

Do you know what most sales reps aren't thinking about when they're engaging with decision makers? They're not thinking about what the decision maker is thinking about. A decision maker's state of mind further complicates outbound sales. This isn't something that a lot of sales trainers & leaders talk about but I think it's a critical part of being successful with outbound. We tend to forget decision makers are real people and far from perfect. They can be crabby, non-attentive, dismissive and even flat out rude to sales people. It's because we're annoying to them. They're not really that person or typically in that type of mood. In most cases, it's only the interaction with sales people that pushes them into that state of mind. This dynamic is one of the main reasons it's so difficult to succeed with outbound sales. In that moment when the decision maker's mood is off, the best outbound sales people are really good at saying the right words. They're really good at crafting the right email. They come off as natural and sincere in their videos. They're great at being bigger than the decision maker's mood. No matter what's going on they usually get their message across.

It's probably safe to say that when most sales leaders and reps lose their jobs, it's because they didn't produce results with outbound sales. This alone makes outbound a hazardous arena for the brave people involved. Adding to the volatility, outbound sales initiatives are usually a company's backup plan. When today's business leaders see their success with inbound marketing dry up, they will often look to outbound sales to fill the void. Some companies go as far as hiring an entire sales team before they've even identified an effective outbound strategy. They usually end up disappointed and the new employees can suffer the consequences. Some situations are more difficult than others but at its core; outbound sales is complicated and many companies fail with it. In order to see results, there usually needs to be these five things:

1. A viable market.
2. Inherent value in the product or service.
3. An effective channel (medium) strategy.
4. Tested sales messaging.
5. On-going development.

One of the most difficult things to do when joining a company is estimating the viability of the market. Part of the CEO's job is to portray a healthy market for investors and employees. This company line usually trickles down to the sales leaders and hiring managers. In reality, some markets aren't viable enough for dedicated outbound sales people. They're either highly mature or what I call 'phantom', mostly a concoction of guesses by business leaders that don't pan out. The outbound sales people who're expected to get results in these types of markets are in a difficult situation. Quotas are usually unrealistic which then forces sales leaders to increase KPIs until eventually the whole thing blows up. Some companies go as far as repeating the process with different personnel only to get the same result. If you find yourself in one of these situations, it isn't the worst thing that ever happened to you. Challenges like this can be an incredible development experience for an outbound sales person. You'll be more savvy and prepared to knock down doors in your next position.

Value can be an illusion for a lot of people in sales and business. What does it really mean? One of the easiest ways to identify your company's value is to understand what prevents the current clients from leaving. Other than cost of change, what are the reasons your current customers continue to choose you? Understanding your company's value is crucial because it's the ammunition outbound sales people need to get decision makers interested in change. Many sales people can go months or even years into their jobs without really understanding the true value their company offers.

This can happen because some companies do a poor job of teaching it to their sales reps. Instead they train everyone on the company lines like, "Our patented technology streamlines all of your marketing activities" or "We have an award winning design team that will enhance your site's visitor experience". When you're in the live action of selling to a decision maker, statements like this mean nothing. An example of selling value would be more like this:

"We can reduce your costs by streamlining your marketing activities."

or

"We can increase your site's conversion by optimizing the design."

Effective outbound channel strategies are one of the most difficult things to discover in business right now. The channel mix that usually works best is the one that fits with the company's messaging, target buying persona and sales personnel. These three factors are going to dictate almost everything in an outbound sales strategy. The best place to start is to identify what channels could potentially reach your buying persona. Some decision makers refuse to answer cold calls, some decision makers never login to LinkedIn, some decision makers will never take the time to read your email. If you can identify the optimum channel mix for your buying persona you'll be in a position to succeed with outbound sales. Later in the book I'm going to breakdown some of the best outbound channels available today.

I think messaging is probably the most overlooked aspect of outbound sales strategies. In my opinion, many reps are using sales messaging that will never interest their target buying persona, unless of course they happen to be looking for that very thing. Sales messaging should be created and perfected by the people actually talking to prospects. They're the only ones in a position to truly understand what's resonating with decision makers. The great thing about sales messaging is that you can continually test out new material. Since most outbound reps make numerous sales touches, there's always an opportunity to run some A/B testing and find out what's working best. Because a majority of sales people I've encountered are struggling with their company's messaging, chapters 4-6 share some simple methods that will help you create your own sales messaging.

There are many professions where continued development is required. Sales should be one of them. Just like an athlete, a doctor, or a sniper. It's important for sales people to be continually perfecting and enhancing their selling skills. But it's not just the continued self-improvement and tactical learning that's important, it's also the decompression time away from selling. Sales life is full of rejection and problems that need to be solved. It makes the occupation difficult to weather year after year. Regular improvement activities can help reps stay focused and recharge themselves. Even courses or exercises outside of the sales realm can be a great way to refresh the team. Communication is a vital part of the sales profession, so workshops or programs that help reps hone their communication skills can be especially helpful.

Everyone in outbound sales is always looking for the silver bullet. What's funny is silver bullets can actually exist. The specific outbound tactic may not always last forever but reps are finding ways to open doors at record setting rates. In some cases they're performing three to four times better than their closet peer. Of course, there isn't one outbound tactic that's going to connect with every prospect. That's why it's important for outbound sales people to have at least a few different strategies in their tool box. I hope this book helps you find yours.

One last note, throughout the book I've included a series of LinkedIn polls that were conducted during August and September of 2020. The point of these polls is to give you some perspective of what real sales people are actually doing.

How many dials they're making, how many emails they're sending, who they're calling on, how they're using LinkedIn to open up doors and much more. By the end of this book you should have a clear picture of what's going on in outbound sales. Did you ever wonder how many cold emails most sales people are sending every day? The answer is below.

LinkedIn Poll for Cold Email Volume – Sept 2020

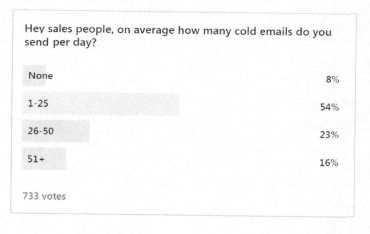

Hey sales people, on average how many cold emails do you send per day?

None	8%
1-25	54%
26-50	23%
51+	16%

733 votes

2

State of Outbound Sales

Crowded and desperate would be one way to describe the state of outbound sales in 2020. There are millions of sales reps all over the world being held accountable for lofty KPIs (Key Performance Indicators). These KPIs typically require reps to contact decision makers repeatedly, over short periods of time. This has created a constant flow of disruptions that has made it really easy for them to ignore all sales people. For decision makers, it's all a blur and they usually put up a wall so they don't have to spend any time thinking about it. In order to find what they need they'll typically rely on internal resources, industry colleagues or research it themselves. Since there are so many companies and reps failing with the targeting aspect of outbound sales, the automatic assumption for most decision makers is that they don't need what you're selling. They believe that you just optimistically think they do. In many cases they're right, further increasing their confidence in resisting.

On top of all this, a lot of the sales messaging being thrown at decision makers is a complete disaster. It's mostly comprised of long value pitches full of buzzwords and company language that's virtually meaningless to decision makers. The programmatic training that sales people receive when it comes to value pitches is working against them. The biggest reason is because forty five seconds of hard core pitch doesn't create a conversation. It's doesn't develop the dialogue you need to effectively communicate with decision makers.

Since decision makers are getting hit up by so many sales people, to be successful you have to stand out. If you're sending the same email as everyone else then it's unlikely you'll get noticed. The most common way reps are trying to stand out is by personalizing their outreach. Unfortunately, many are also failing with this. Most "personalization" is ineffective because the reps aren't digging deep enough into the prospect. It's probably because they don't have much time. Since the number one objective is to meet their KPIs, doing the things necessary to succeed can become secondary. Instead, they're scratching the surface and sending personalized messages that aren't personal. For example, I get this opening statement constantly in cold emails and LinkedIn messages:

"Hi Matt, I see you're doing great things at Subroot."

I usually think to myself, really you see that I'm doing great things, I'm not sure I believe you. Personalization is more like this:

"Hi Matt, I recently read your book and really enjoyed it."

Is this manipulation? Probably. But when it comes to outbound sales, things like this work. Your personalization has to go deeper than everyone else's if you want to standout. There's much more on this in Chapter 8.

What do you think is a good response rate for cold outreach? If a rep could hit 10% that would be incredible, even 5% would be really good. The average response rate for cold outreach is probably less than 1% and for many outbound reps it could be less than .1% (1 in 1000). But you know what? They still have to keep pushing and completing their KPIs. It's likely they'll also need to continue using the company's plan, even if it's not working. Many outbound sales people don't have the autonomy to try new channels or different messaging. Companies usually take this stance because the strategy worked at one time. These circumstances are common ground on the outbound sales battle field.

Another hot topic when it comes to the state of outbound sales is prospect contact data. You can spend an arm and a leg paying for accurate contact information and many companies choose to do so. When it comes to contact data there are generally two types of providers, human verified and not human verified. The data providers giving you the latter are constantly scrapping the internet for available contact information. These services can be pretty inaccurate but the affordability and ease of use make them something to consider. The human verified data is much more accurate and expensive. The providers typically guarantee around 95%

accuracy on email addresses and phone numbers. Obviously the more accurate the contact information the more productive the sales team, every sales org or person needs to decide for themselves where they want to be with data accuracy and costs. Another option is to check out fastpeoplesearch.com. It's completely free and provides reasonably accurate information per Scott Koethe, the greatest commercial realtor in Southern California. Dude has his own calendar.

There's also another kind of data, most people would probably call it analytics. The data I'm talking about includes things like open & response rates on emails, videos and social media messages, cold calling conversion numbers and associated analytics like length of call, rep vs. decision maker talk time, etc. There are companies like Gong and Chorus that provide some really interesting data surrounding these analytics. This information can be very helpful when framing an outbound sales strategy. Keep in mind, this data doesn't break down which decision maker is being targeted, for example Vice President vs. Manager. In my experience, this one variable can make an enormous difference.

As you probably know there is sales tech available that automates the outbound sales process. Sales engagement technology allows sales people to manage cadences and complete tasks like sending emails or dialing phone numbers. This type of automation makes it much easier for reps to manage a large number of prospects and sales touches. If you're manually dialing the phone fifty times a day (I run into a lot of people doing this) you may want to check out sales engagement tech with an autodialer.

It may be safe to argue that outbound sales is getting both harder and easier at the same time. New competition is being continuously created and decision makers have more ways to buy things without a salesperson. But new mediums like social media and video messaging have given reps better ways to get in front of prospects. There is one thing that I think will always hold true in outbound. Your ultimate results will be determined by how often you're able to push through the first layer of rejection. That's the topic of the next chapter. Just in in case you were wondering, opening up doors is still the hardest part of sales!

LinkedIn Poll –The most difficult part of sales – Sept 2020

Hey Sales People, which of these is the most difficult part of your job?

Opening up doors	55%
Closing business	11%
Fulfilling KPIs	13%
Dealing with management	21%

423 votes

3

First Layer of Rejection

One of the real reasons outbound sales is so difficult isn't something you hear too much about. When you break down the process of entering a new account; the hardest point to overcome is actually before you've even made contact. No matter the medium, going from an unknown to a point where the decision maker is attempting to understand your message, that's the "first layer of rejection".

The reason I'm bringing this up is because the first layer is where most sales touches are eliminated by decision makers. Since the first layer of rejection comes before the decision maker even understands your message, you're not really being rejected on merit. Instead, they're hastily rejecting the idea of even trying to understand the information. Maybe they don't have enough time to consume all the sales messages being thrown their way every day. Decision makers are using the little information available to quickly decide if they should absorb more. If there's any doubt, they deem the outreach unimportant and move on with their day.

The first layer of rejection is a point in time when the decision maker still sees you as a bothersome rep selling something they don't need. It's the time when you ask for 28 seconds to tell them why you're calling and they tell you to piss off. It's when they mark your email as spam before they've even read it. It's when they tell you "we're not interested" before you can say your name.

The first layer is an easy place for decision makers to end things because then they don't have to spend any time thinking about you or what you're selling. They don't have to commit to an uncomfortable conversation with a pushy sales rep or entertain the likelihood of rejecting you later in the sales process. Even more importantly, they don't have to consider making a scary change.

Until you get past the first layer of rejection with a decision maker, you haven't gotten anywhere. They don't know you. More importantly they have no idea what you sell or how it can help them. You're an unknown and they're going to do their best to keep it that way. Once you get past the first layer of rejection, your company's value and your ability to convey it will determine if there's a next step. But this first layer is where most outbound sales campaigns end. Once you figure out how to consistently push through it, you will be a master of outbound sales.

So, how do you shoot through the first layer of rejection? Well, that's why I wrote this book. I'm going to tell you everything I know, which includes providing you with some simple ways to create your own sales messaging. It doesn't matter if your outbound tactics are great, if your sales messaging sucks. Every outbound sales strategy should start with creating effective messaging and it should be continually refined and tested. In my 'Sales Conversation Superstars' on-line program I teach reps how to create their own sales messaging. In the next three chapters I've included all of our little tricks to help you find an optimum purpose for your outreach, create great questions for decision makers and turn around the toughest objections.

4

Nail your purpose

One of the easiest ways for decision makers to reject outbound sales people is when they don't identify with the purpose of their outreach. Whether it's a conversation, written message, voicemail or video message, there is usually one question on the mind of the decision maker, "Why are you contacting me"?

Decision makers are usually listening or scanning for this single piece of information so they can eliminate the sales person and return to whatever they were doing. If you're unable to connect with your purpose there's a good chance you've lost their attention.

-If you say you're calling about printing services, they already have a printer. Goodbye.

-If you say you're reaching out to them about sales engagement tech, they already have a vendor. Adios.

-If you say you're contacting them about consulting services, they don't need a consultant. Goodbye.

As an outbound salesperson the odds are stacked against you because decision makers are looking for WHY NOT. This makes finding an optimum purpose critical when trying to push through the first layer of rejection.

What does optimum purpose mean? It's a purpose that takes the decision maker from 0 to 1 on the interested meter. Even more importantly it keeps the decision maker's attention. If it falls into a category where they already feel satisfied, they're more likely to dismiss your outreach.

If your purpose is something that lands on a problem or an area of improvement for the decision maker, they'll be more inclined to engage. The difference between a purpose that captures their attention and one that doesn't can be subtle. This is why outbound sales is a game of inches. If you're able to exchange a few of the right words to the right people at the right time, you might just open up a huge door.

 Patrick Moss · 12:36 PM

Hey Matt! going well. I set two big meetings with the script we developed. I've been in workshops and vacation so I havent had a chance to get to our second session.

Here's an easy method to create an optimum purpose for your cold outreach. At the end of the chapter I've included a workbook so you can get started creating yours.

Creating an Optimum Purpose

First step is to identify:

1. Who or what benefits from the product or service?

2. What is the benefit?

Second step is to COMBINE them to make your purpose.

Optimum Purpose Examples

If you sell <u>Sales Engagement Technology</u>:

Who/what benefits: Sales team.

What's the benefit: Increased productivity.

Possible Purpose: *"I'm reaching out about your sales team's productivity."*

If you sell <u>Car Insurance</u>:

Who/what benefits: Your car.

What's the benefit: Protection.

Possible Purpose: *"I'm calling about your car's protection."*

If you sell <u>Accounting Services</u>:

Who/what benefits: Your taxes.

What is the benefit: Maximum return.

Possible Purpose: *"I'm calling about a maximum return on your taxes."*

If you sell <u>Contact Data</u>:

Who/what benefits: Sales team.

What is the benefit: Increased efficiency with accurate data.

Possible Purpose: *"I'm emailing about your sales team's efficiency.*

If you sell <u>Network Security Software</u>:

Who/what benefits: Your network.

What's the benefit: More security.

Possible Purpose: *"I'm calling about more security for your network"*

If you sell <u>Web Development</u>:

Who/what benefits: Your website.

What's the benefit: Conversion.

Possible Purpose: *"I'm contacting you about your website's conversion."*

If you sell <u>Cleaning Services</u>:

Who/what benefits: Your facility.

What's the benefit: Maintaining cleanliness.

Possible Purpose: *"I'm calling about maintaining your facility."*

If you sell <u>Sales Training;</u>

Who/what benefits: Sales team.

What's the benefit: Increasing sales numbers.

Possible Purpose: *"I'm contacting you about increasing your sales team's numbers."*

Optimum Purpose Workbook

Who or what benefits from what your company provides?

```
┌─────────────────────────────────────┐
│                                     │
└─────────────────────────────────────┘
```

What is the benefit?

```
┌─────────────────────────────────────┐
│                                     │
└─────────────────────────────────────┘
```

Optimum purpose:

```
┌─────────────────────────────────────┐
│                                     │
│   I'm contacting you about...       │
│                                     │
│   ─────────────────────────────     │
│                                     │
│   ─────────────────────────────     │
│                                     │
└─────────────────────────────────────┘
```

Only use a No. 2 pencil when completing this workbook.
(I'm totally kidding, use whatever you want, big seller!)

5

Creating Good Questions for Decision Makers

My entire career I've been on the frontlines selling, first as a sales rep and sales manager for a major corporation and then selling for my own companies. I've continually found conversations to be the best and sometimes the only way to push a sale forward. Developing communication skills is one of the most important things you can do as a salesperson. One of the keys to having successful sales interactions is being able to ask decision makers the right questions. In this chapter I share a simple framework that I developed to help sales people create their own questions. By the end of this chapter you'll be ready to create some good questions for your decision makers.

As you have probably already experienced, initial sales conversations can be really difficult. One of the reasons they're hard is because of the person on the other end. Normally this is someone who doesn't want to hear what you have to say. In many cases getting the decision maker to begin having a normal back and forth conversation with you, is the key to having a meaningful exchange. This is why the questions you ask are so important, especially the first two or three.

 Suzi Fornoff · 2:28 PM

FYI, I just got off the phone with a prospect, using your questions and the little boost of confidence you gave me I turned a cold call into a twenty minute conversation and a meeting in January.

-Thank You

I think most reps have already experienced this while talking with decision makers; sometimes it's just a single question that changes everything. Here's what can be accomplished with the first few questions you ask a decision maker:

1. Pertinent questions can help bring the decision maker into the conversation.
2. A pointed question can reveal that there's something the decision maker doesn't know.
3. The right question can help deflate the agitation a decision maker might be feeling.
4. Magnifying questions can help expose the problems your company solves.

Avoid Comfort Questions:

Before I start creating questions I'm going to cover a few of the questions you should avoid. These questions probably apply more to cold calls but they're something to keep in mind whenever you're talking to decision makers for the first time. "Comfort questions" are what most sales people ask when they're feeling the most uncomfortable during a sales conversation. The most popular comfort question, which I'm sure you've heard, is "How are you doing today?" Another is when the decision maker answers the phone with, "Hi this is Rick", and the rep immediately asks, "Is Rick available?" Asking comfort questions isn't the end of the world but they can put you in a negative position to start. The biggest problem with these questions is they don't take us where we want to go, which is telling the decision maker how we can help their company. Here are some other reasons to avoid comfort questions:

1. Asking a complete stranger how they're doing is always going to get mixed results. Decision makers are busy and the last thing they want to answer is a meaningless question.
2. Questions in a sales conversation are like bullets. You're only going to get so many chances to get the decision maker engaged. You don't want to waste any on personal or irrelevant topics.
3. The whole point of talking to a decision maker is to let them know how your company can help. The questions you pose should take the conversation in that direction.

Asking for Permission

Another popular strategy is to ask a decision maker for their permission to tell them why you're calling. Even though this is a respectful approach, it doesn't engage the decision maker and in the end can make it easier for them to leave the conversation. Here are some other reasons to avoid asking for permission:

1. You're giving decision makers a chance to leave at the very beginning of the call. Some of them will take you up on that. Others might feel like all they need to do is to hear your pitch and then they can leave the conversation.
2. By asking for an opportunity to pitch, you make that the deciding factor. If you don't connect on something pertinent for the decision maker, the interaction is usually over.
3. Since you're not developing rapport, getting the decision maker interested can be an uphill battle. It's unlikely they'll consume much of the information you're trying to communicate.

Closed-ended vs. Open-ended Questions

People in sales are often taught to ask open-ended questions. Though this may seem like a logical approach, it's not the best way to begin an initial sales conversation. Closed-ended questions give you a much better chance at creating a dialogue with a decision maker. Here's why:

1. The first seconds of any sales interaction can feel awkward for a decision maker. By asking them a closed-ended question it makes it easier for them to respond and engage in the conversation.

2. Closed-ended questions make it easy to be prepared for the decision maker's response. Especially if you're asking a Yes or No question.

3. One objective during the conversation is to share some key value information. Closed-ended questions can help guide the conversation in that direction.

Creating Questions

The questions created from this exercise aren't just for cold calls. They're also effective questions to ask on inbound follow ups, warm calls, or when you're working the floor at trade shows. These questions can even be used in your cold emails and other messages.

I'll be using four real life examples in order to create some sample questions. I think you'll find this to be a pretty straightforward process. The first step is to identify what problems your company solves (yes, every product or service solves some type of problem, even yours!). The second step is to identify how your product or service solves each problem. In this first example we're selling Sales Engagement Technology.

Example 1: Sales Engagement Technology

What problems does a sales engagement technology solve?

Problem 1: Low activity for sales reps.
Problem 2: Maintaining lead accuracy.
Problem 3: No management visibility.

Okay, right now you might be thinking the problems you solve are the questions. Well not exactly. We all know you wouldn't want to ask a grumpy decision maker if their reps are having a problem with low activity. That probably wouldn't create the type of exchange we want. So, before we can create our questions we need to identify how the technology specifically solves each of these problems.

How does sales engagement technology solve these problems?

Problem 1: Low activity for sales reps.
Solution 1: Increases the number of sales touches.

Problem 2: Keeping track of prospects.
Solution 2: Provides simple data management.

Problem 3: No management visibility.
Solution 3: Offers full management visibility.

Forming Questions

It's important to understand that there will be a few different questions that you can create from this information. There's usually more than one right answer but there are also wrong questions. You always want to make sure you're asking questions the decision maker will answer.

Recapping the first problem and solution:

Problem: Low activity for sales reps.
Solution: Increase the number of sales touches.

Possible First Question: *"Do you think your team is making enough sales touches?"*

Recapping the second problem and solution:

Problem: Keeping track of prospects.
Solution: Provides simple data management.

Possible Second Question: *"Do your reps have a hard time keeping track of prospects?"*

Recapping the third problem and solution:

Problem: No management visibility.
Solution: Offers full management visibility.

Possible Third Question: *"Do you have visibility to what your reps are doing every day?"*

Example 2: National Printing Company

In my second example, I'll be covering a National Printing Company. To give you some background, this company has twenty printing facilities located throughout the US. They offer a standard 3 day turnaround on most orders and have a sophisticated order system that allows their customers to send, change, and view the status of orders in real time. As you know, the first thing we need to do is identify what problems this National Printing Company solves.

What problems does this National Printing Company solve?

> Problem 1: Long turnaround.
> Problem 2: High shipping costs.
> Problem 3: Order visibility.

How does the National Printing Company solve these problems?

> Problem 1: Long turnaround.
> Solution 1: Provides a standard three day service.
>
> Problem 2: High shipping costs.
> Solution 2: Multiple facilities throughout the country.
>
> Problem 3: Order visibility.
> Solution 3: Systems provide full visibility of orders.

Forming Questions

Recapping the first problem and solution:

Problem: Long turnaround.
Solution: Provides a standard three day service.

Possible First Question: *"Do you run into any projects where you need a quick turnaround?"*

Recapping the second problem and solution:

Problem: High shipping costs.
Solution: Multiple facilities located across the country.

Possible Second Question: *"Would you like to reduce your shipping costs?"*

Recapping the third problem and solution:

Problem: Order visibility.
Solution: Systems provide full visibility of orders.

Possible Third Question: *"Do you have any visibility after you place your print orders?"*

Example 3: Unified Communications Provider (UCP)

In this example, I'll be creating questions for a Unified Communications Provider. This company provides a variety of communication services including phone, cellular, internet, and conferencing. Because they offer a menu of services their customers are able to consolidate communication vendors. They also have the option to easily integrate UCP's communication system with their other business systems.

What problems does this Unified Communications Provider solve?

> Problem 1: Having to manage multiple vendors.
> Problem 2: Communication systems don't integrate.
> Problem 3: No options for remote workers.

How does the Unified Communications provider solve these problems?

> Problem 1: Having to manage multiple vendors.
> Solution 1: Provides everything in one platform.
>
> Problem 2: Communication systems don't integrate.
> Solution 2: Flexible integrations.
>
> Problem 3: No options for remote workers.
> Solution 3: Several options for remote workers.

Forming Questions

<u>Recapping the first problem and solution:</u>

Problem: Having to manage multiple vendors.
Solution: Provides everything in one platform.

Possible First Question: *"Are you using multiple communication providers?"*

<u>Recapping the second problem and solution:</u>

Problem: Communication's systems don't integrate with other business systems.
Solution: Flexible integrations with most business systems.

Possible Second Question: *"Does your communication system integrate with your other business systems?"*

<u>Recapping third problem and solution:</u>

Problem: No communication options for remote workers,
Solution: Several options for remote workforce.

Possible Third Question: *"Do you have employees working in the field or remote?"*

Example 4: Sales Training Company

In the final example, I'll be creating questions for a Sales Training Company. This company provides a variety of training services, including general sales, cold calling, and technology training. Their expertise in these areas ensures sales people are properly trained and prepared for the field.

What problems does this Sales Training Company solve?

> Problem 1: Low sales numbers.
> Problem 2: Converting phone conversations.
> Problem 3: Technology adoption.

How does the Sales Training Company solve these problems?

> Problem 1: Low sales numbers.
> Solution 1: Optimize outbound channels.
>
> Problem 2: Converting phone conversations.
> Solution 2: Refine sales messaging for dialogue
>
> Problem 3: Technology adoption.
> Solution 3: Expert level training on sales tech stack.

Forming Questions

Recapping the first problem and solution:

Problem: Low sales numbers.
Solution: Optimize outbound channels.

Possible First Question: *"Do you feel like your sales team is using the right outbound channels?"*

Recapping the second problem and solution:

Problem: Converting phone conversations.
Solution: Refine sales messaging for dialogue.

Possible Second Question: *"Do you think your team is having good phone conversations?"*

Recapping the third problem and solution:

Problem: Technology adoption.
Solution: Expert level training on sales tech stack.

Possible Third Question: *"Are you having any problems getting your reps to adopt new technologies?"*

After the Question

You're probably wondering what happens after the decision maker responds. Typically, there are only two responses when you're asking a decision maker closed-ended questions. One is what I call the "positive response", like: *"Yes, we're having problems getting our reps to adopt new technology."* The other is the "negative response", like: *"No, our reps adapt to any new tech very quickly."*

Whether the positive response is a Yes or a No depends on the question you ask. With some questions, the positive response could be a No.

Positive Response

The positive answer gives us a great opportunity to reply with a mini-pitch, followed up with another question. Like this:

Rep: *"Are you having any problems getting your reps to adopt new technologies?"*

Decision Maker: *"Yes, we're having problems."*

Rep: *"I hear that quite a bit, we provide tech training for sales teams. It helps get reps up and running really quickly. Can I also ask, do you feel like the team is having good phone conversations?"*

Negative Response

If the decision maker gives you the negative response there are two different ways to respond. One is with an open-ended question that further qualifies their response.

Decision Maker: *"No, they adapt well."*

Rep: *"It's great you're doing well with that, I don't hear that often. Are you guys doing anything specific that helps them adjust quickly?"*

The second option is to acknowledge their answer and move on to the next question. Like:

Decision Maker: *"No, they adapt well."*

Rep: *"That's great you're doing well with that. Can I also ask, do you feel like the team is having good phone conversations?"*

Tone

Tone is the one thing that I can't give you any perspective on with this book. If you felt like some of the above questions or statements wouldn't work, you might change your mind if you heard them with the right tone. It's an important part of communication, especially in a selling situation. It's very important to be aware of your own tone. If you need to work on it (many people do), start listening to the best reps on the team talking to prospects and customers. You can even try and get recordings. Then mimic their tone in the mirror. Keep working on it until it feels more comfortable.

Creating Questions Workbook

What problems does your company solve?

1.

2.

3.

How does your company specifically solve these problems?

1.

2.

3.

If you have a highlighter handy, go ahead and highlight the main point from each problem or solution, like in my examples on Page 41(outbound channels, phone conversations and technology adoption).

Now create your questions. Start the sentences with things like: *Are you having any problems...? Do you think your team...? Are you using...? Do you have...? Do you feel like...?*

1.

2.

3.

PS. If you happened to botch these, no worries! There are more copies at the end of the book or on our website at subroot.com/workbooks.

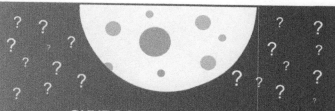

SKYROCKET YOUR SALES
BY ASKING QUESTIONS

Have you ever wondered what makes the top sales people the best? This graphic shows why sales people who become really good at asking questions are usually the top performers. Of course, the quality of the questions and the way in which they're asked will ultimately determine the results.

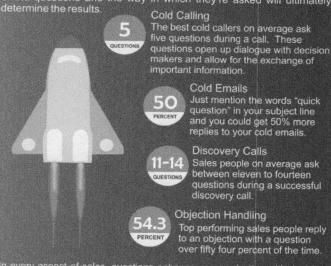

5
QUESTIONS

Cold Calling
The best cold callers on average ask five questions during a call. These questions open up dialogue with decision makers and allow for the exchange of important information.

50
PERCENT

Cold Emails
Just mention the words "quick question" in your subject line and you could get 50% more replies to your cold emails.

11-14
QUESTIONS

Discovery Calls
Sales people on average ask between eleven to fourteen questions during a successful discovery call.

54.3
PERCENT

Objection Handling
Top performing sales people reply to an objection with a question over fifty four percent of the time.

In every aspect of sales, questions enhance communication with decision makers and give you the greatest chance to effectively communicate your company's value. Figure out the right questions and your sales numbers should Skyrocket!

Sources:
Chorus - State of Conversation Intelligence 2020
Sales Loft - n:phrase used 59,640 times sample of 6 million emails.
Gong - The 7 Best Discovery Call Tips
Gong - Chris Orlob article - What separates crazy successful salespeople from everyone else.

6

Handling Tough Objections

I've been handling objections from decision makers for over twenty years. At this point, I've probably encountered them all. (And as a decision maker for my businesses, I've thrown quite a few of my own at sales people.) Objections can be the best thing that happens to a sales conversation. In some cases the objection allows the salesperson to disqualify a prospect because they're not a good fit. Other times an objection gives you a great opportunity to convey details that are going to get the decision maker interested in making a change. The reality is, anytime you open a door you're probably going to face some objections. When you're able to move past tough objections it can be the turning point. This makes your ability to handle objections an important part of being successful with outbound sales. In this chapter we're going to dive into some of the most difficult objections sales people face and provide you with some different ways to handle each one.

Most sales conversations end with an objection that the sales person is unable to overcome. Pushing through that objection is the only way to advance the conversation. The most common objections are usually thrown at sales people during the first layer of rejection, before the decision maker has very much information. The toughest objections are what I describe as absolute. For example, "We're not interested!" This is probably the most common objection on the planet.

Absolute objections are difficult to respond to because the decision maker is usually delivering them with a definitive tone. Because in their mind, with all of the information they currently have, they're truly not interested. But you know what? They probably don't have all the information. It's your job as a salesperson to make sure they have a clear picture of what your company can do for them.

When you're facing a tough objection there's usually just a single chance to say something that will bring the decision maker back into the conversation. You'll only get a brief moment to say words that satisfy or deflect the objection. It's a difficult spot to be in. One of the best ways to handle it is by changing direction and asking the decision maker a question of your own. The question should be related to the objection and it should try to take the decision maker away from their current line of thinking.

One of the big problems I see with canned responses for objections, they're just not realistic. When a decision maker throws out a tough objection the rep doesn't have time to spout out an entire paragraph of information. The response needs to be short because you don't have much time. That's why it's also important to engage the decision maker quickly, because in their mind the conversation is over. You have to recapture their attention before you can get your message across.

There's a reason they're called common objections. Sales people from all over the world in every industry are encountering these same objections from decision makers. In the Smart Outbound program we create messaging to help reps handle their most common objections. The remainder of this chapter includes some responses to the most common objections that we see in our program. You may need to adjust these responses to fit your specific situation. Feel free to insert some of the questions you created in the previous chapter when you can.

Keep in mind, one of the most important parts about handling objections is your tone. As I mentioned earlier, you can always improve your tone with practice and repetition.

Common Objections

"We're not interested."

This is the most common objection sales reps encounter and it can be very difficult to overcome. One reason is because decision makers may feel disrespected if you don't heed their objection. If you give them the wrong response, they will often come back at you with an even stronger, "I told you we're not interested". Here are a few different ways to handle this very common objection. Possible Responses:

a. *Hey Ron, do you even know what my company does?*

b. *Do you even know what we do, Ronda? Some people have us in the wrong category.*

c. *Ricardo, do you understand some of the things that our tech provides? I think you might find it interesting.*

d. *Rana, I haven't even shared the most important information. Can I have a second to explain?*

e. *Ramesh, it's my job to make sure you have all the information. Did you know we work with (COMPETITOR) REFERNCE?*

f. *Are you interested in lowering your costs, Will? Because we could do that for you.*

"We're all set with our current provider."

This is another really common objection and it falls right in line with "We're not interested". The decision maker is usually thinking, "If it isn't broke, don't fix it". The opportunity here is that in many situations, it is broken or it could be a lot better. The decision maker may not be aware of this. Our response to their objection gives us an opportunity to point that out. Here are some ways you can handle "We're all set". Possible Responses:

    a.   *I know you have a vendor in place. Do you understand how our product/service is different?*

    b.   *I know you have a vendor in place. Did you know we usually save our clients 20%?*

    c.   *Most of our clients felt that way. What we offer is different than your current situation. Did you know we can help DO THIS GREAT THING?*

    d.   *I know you already have a source. We could help make sure you're getting a good deal. Did you know we can DO THIS GREAT THING?*

    e.   *I understand you have something in place. Would it be helpful to see what else is available? A lot of things are changing in the industry.*

"You've reached me at a bad time."

This is a tough objection because you want to be respectful of the decision maker's time. But the reality is if you let this person go, it's unlikely you'll speak to them again. It's always a "bad time" for decision makers, however there's a chance that this really IS a bad time. When obliging this request it's important to make it clear your pursuit will continue. This can snap the decision maker back into the conversation because they might want things concluded right now. Since we must be respectful of their time, it leaves us with only a few different responses to this objection. Here are a couple ways to handle "You've reached me at a bad time". Possible Responses:

a. *Sorry Ron, what I have to share only takes a minute but no problem. When is a better time I can reach you?*

b. *Sorry it always is, Ronda. When is a better time to reach you? I want to share some info you might find valuable.*

c. *Oh I'm sorry, Rana. When is a better time I can reach you? I'm calling about something that might interest you.*

"We don't have any budget."

This is another classic objection that most outbound sales people face on a regular basis. We're contacting them out of the blue. Of course we're not part of their preplanned budget. Some decision makers will use this objection as a way to put sales people off rather than rejecting them altogether. What you need to know about this objection is that when things make sense, budgets can change. Maybe what you offer is going to reduce their employee's workload, increase their overall sales, lower operating costs, or improve efficiency. All these things could warrant an immediate change if the business leaders believe it's beneficial for their company. With this in mind, here are a few different ways to handle, "We don't have any budget". Possible Responses:

    a.   *Ron, what we provide usually isn't part of the plan but it can affect other areas. Did you know we can do this GREAT THING?*

    b.   *Our technology can impact other parts of your budget, Ronda. Did you know we can DO THIS GREAT THING?*

    c.   *I understand it's probably not your budgeting time. Did you know we're usually able to SOLVE THIS PROBLEM?*

"I'm not the decision maker."

This objection is a little easier to believe because decision makers are unlikely to mislead sales people in this regard. They're more inclined to tell you "No". Even though the person you've connected with isn't the direct decision maker, once you start sharing information about your company their areas of responsibilities could be impacted. They could quickly become an advocate for you. Here are a couple ways to handle "I'm not the decision maker". Possible Responses:

a. *That's okay, I'm just looking for information. Do you know if your company is having any problems with PROBLEM YOU SOLVE?*

b. *Sorry, I wasn't sure if you were. Do you know how a decision like this would be made at your company?*

"Send me some information."

This objection can be like the kiss of death. When decision makers use this objection they're usually leaving the conversation and they don't plan on talking to you ever again. This thread of hope they're giving you by asking you to send them more information, is their ticket out the door. Most decision makers throw out this objection when they can't connect what you provide, to what they might need. With this in mind, here are a few ways to try to tackle, "Send me some information". Possible Responses:

a. *Okay, we have a lot of different materials. Is there something in particular you want to see or an area you're trying to improve, Ron?*

b. *I don't have much to send, Rita. The information I need to share really takes an explanation. Are you having any problems with the PROBLEM YOU SOLVE?*

c. *Well, I'm probably your best source of info. Could I just quickly explain one of the ways we're benefiting our clients?*

7

Outbound Options & Cadences

The popularity of social media sites like LinkedIn, combined with the ability to send video messages has given sales people more outbound options than ever before. Combining these new channels with cold calling, cold emailing, voicemails and direct mail gives sales people several ways to get in front of decision makers. These new mediums are timely because it's incredibly congested trying to reach prospects. But all these options have also made outbound sales more complex. Finding the right channel mix for your target persona has become a lot like cracking a secret code. If you can find a repeatable outbound sales strategy, you'll have more control over crushing your quota.

These new channels have also created a larger gap between the best performers and the rest of the pack. Reps who find hyper-productive outbound tactics can outperform their average peer by 3 to 4 times. It's because they've found repeatable tactics that allow them to penetrate accounts at a much faster rate. Sometimes these strategies can be repeated by other reps on the team, sometimes they can't. My attitude when it comes to any outbound sales strategy that someone claims to be effective is, "let's give it a shot". Outbound sales should be more R&D than "follow this old protocol". Continual success is hard and the more tactics you have at your disposal, the better.

As you probably know, being creative has become an important part of succeeding with outbound sales. That creativity can come in the form of personalization, e-mail copy, direct mail, a cadence, you name it. Today's outbound options give you a lot of ways to be creative but you're also competing with other sales people who in their own way are trying to standout. Once you're able to standout you've essentially broken through the first layer of rejection.

The reality of outbound is that most decision makers aren't going to pick up your cold call. Most decision makers will pass on reading your email. Most decision makers will never watch your video (except for LI video messages). Most decision makers aren't active on social media (yet). For these simple reasons, having a multi-channel outbound strategy is going to help you open up MORE doors. Finding the mediums that work best for you, is a whole other ball game.

LinkedIn Poll on Number of Outbound Sales Mediums – Sept 2020

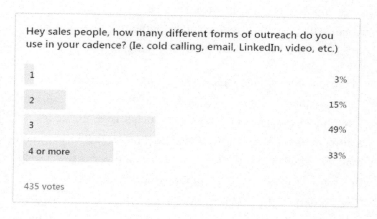

Hey sales people, how many different forms of outreach do you use in your cadence? (Ie. cold calling, email, LinkedIn, video, etc.)

1	3%
2	15%
3	49%
4 or more	33%

435 votes

Cadences

You can't talk about outbound sales right now and not talk about sequences or cadences. For those of you who haven't been around a cadence, it's a set of preplanned sales touches that usually include phone calls, emails and social media messages. Cadences have quickly become one of the hottest topics in sales. One of the most asked question in outbound right now is, "What cadence is working for you?"

One of the great things about cadences is that they keep you organized. Your CRM or engagement tool tells you to do this task, on this day, which makes it really easy to maintain your outreach. The downside is you're completing dozens of tasks every day and they begin to feel like just that, tasks. In reality each one of these sales touches is a game winning free throw and having that attitude is what will help you be successful. It's really hard to bring passion to forty, sixty, or even 100 sales touches on any given day. This is one of the many reasons that people involved with outbound sales are pretty special. (Don't let it go to your head though, jk.)

There is one thing which most sales cadences have in common that is working against all of them. They all end. The typical cadence is 5 to 8 sales touches over the course of a few weeks. You know who else knows that? Decision makers! They know if they just wait it out for a few weeks this rep will move on to another prospect with their, 3 emails, 2 voicemails, 2 LinkedIn messages and 1 video. ☺ Here's a LinkedIn poll that will give some perspective on the typical length of a cadence.

LinkedIn Poll on Length of Cadence – Sept 2020

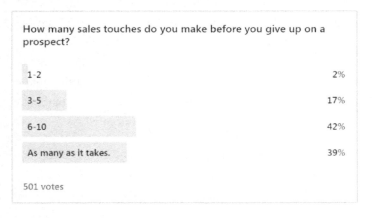

Cadences or sequences are important but they're not going to make you successful in outbound sales. I think it's important to understand that what's working for someone else won't necessarily work well for you. In many sales positions, you have to find your own successful sequence and methods. Since sales cadences dictate every outbound sales touch, they've become an overinflated topic. It's usually the "specific methods" within the cadence that are really driving success.

Most cadences will include a mix of cold calls, cold emails, LinkedIn messages and maybe even direct mail, video messages or email sequences. Your target market, the resources your company offers and a few other factors will determine what type of cadence is best suited for you. Arguably, the most important part of the cadence is the first few sales touches. This is the part of the cadence where you could have the greatest chance to push through the first layer of rejection. Sales touches later in the cadence can be successful but it becomes more and more unlikely as you go. But again, every situation is different and I think it's important to beware of that when you create your cadences.

Example Cadence

Here's one of the cadences I use for Subroot.

	Sun	Mon	Tue	Wed	Thu	Fri	Sat
Week 1	06	07	08	09	📞	11	12
Week 2	13	14	15	📞	17	18	19
Week 3	20	21	22	23	✉ 📞	25	26
Week 4	27	28	29	30	01	02	03
Week 5	04	05	✉	07	📞	09	10
Week 6	11	🔲in	13	14	15	16	17
Week 7	18	19	20	📹	22	23	24
Week 8	25	26	27	📞 ▢	29	30	31

I start this cadence off with two dials. I do this because with a certain percentage of prospects, it's going to end right there (15%). They may get disqualified or they may hit the pipeline but because I made these dials, I didn't spend any time writing a message. Plus, if I'm able to connect with one of these dials I'll have an actual conversation, which beats everything in outbound sales.

If I didn't catch anyone on my first two dials, I usually go to a tactic that I call the "Lure email". The tactic includes an email and phone call on the same day. (I provide complete details on this strategy in Chapter 11.)

After the Lure email I send a longer email giving the decision maker more information on how we can help. Later that week I follow it up with a phone call and sometimes I'll leave a voice message referring to the email. I want them to hear my voice.

If none of that works, I take my talents to LinkedIn and send them an InMail or direct message. If they dodged me on LinkedIn and I still feel the prospect is a good fit, I send them a video message. I keep it short and introduce myself. I reference something personal and maybe recap a little of the email I sent.

After that I make one more dial and depending on the prospect, I may set them up on one of our email sequences. This will send them some content and tools they might find helpful. If it's a really ideal prospect I'll put them on my "Must Reach" list and keep calling periodically for six months. I need to know.

In the remainder of the book I cover a handful of effective outbound sales tactics. Some of these tactics are old and some are new. Some are easy to execute and others aren't. Some might fit your style and others won't. The reason I'm providing an overview of these specific tactics is because they're working. Right now, sales reps out there are using these methods to open up doors at a high rate. They could work just as well for you.

As you read these tactics I would encourage you to try and apply them to what you're selling, even if you don't think it's a good fit. The reason I'm saying this is because experienced sales people tend to pass on tactics they've already attempted or maybe don't feel comfortable doing. If you gave the tactic a shot last month and it wasn't working, I get it. But if it was years ago, it's a much different time right now. It doesn't matter where you're at in your sales career, you can always decide to take on the challenge of being uncomfortable. It might be invigorating.

8

Personalized Emails and Phone Calls

This strategy continues to produce results for reps in almost every industry. The specific cadence isn't as important as the email and though the email needs to be interesting, it should also be well written. By well written, I mean concise. Anytime you write an email you should review it and remove as many words as possible. You might be surprised at the number of unnecessary words you find.

Email

Many sales people who send a lot of cold emails will tell you to keep it short. One reason is because a majority of emails are read on mobile. So a snooze fest of information probably won't be read. (Like most sales books, bazinga!) Short version emails are one or two paragraphs (one to three sentences each) that outline a problem and a possible solution. Ideally, long form emails are no more than three to four paragraphs and will include a personalization, problem identification, proposed solution and credibility statement. Many people think it's best to end your initial email with a soft call to action but this might be something to test out. In the second half of this chapter I provide four cold email templates. These examples are for longer form emails but like many reps, you may decide to send short ones. Use the templates as a starting point and just remove paragraphs accordingly. Here's a LinkedIn poll that breaks down the typical length of cold emails reps are sending.

LinkedIn Poll on Length of Cold Email – Sept 2020

Hey Sales People, on average how long are your cold emails (not including call to action)?

1 paragraph	41%
2 paragraphs	42%
3 paragraphs	16%
4 paragraphs	1%

674 votes

Phone Calls

Once you've sent your initial email it should be followed up with a minimum of three to four phone calls over the next few weeks. If the decision maker initially ignored the email, which is likely, the phone calls may push them to go back and read it. Even if they don't connect with the email's message, calling them can sometimes provoke a response because now they're motivated to get you to stop chasing. (I provide a lot more information on cold calling in chapter 11.)

Next Step

If your dials don't spark a response the next step is to send another short email or leave a voicemail. In your second email you should avoid statements like "pushing this to the top of your inbox" or "circling back". It's better to take a firmer position and request a specific time to talk. This approach is effective because decision makers can be compelled to take action once they know there's a date and time in question. They're motivated to get you to stop professionally pestering them. The content of the email should assume they read your initial email, which can help push them to actually read it.

Hi Ron,

In regards to my previous email, I wanted to see if you're available on Tuesday, Oct 6th at 10:00am for a brief call.

Or let me know if there's a better time for you.

Best regards,
Matt Wanty

Voicemail

If you decide to leave a voicemail, I think the most productive purpose for a voicemail is to push the decision maker to read your original email. The contents of the voicemail should be selling the email. Voicemail Sample:

Hi Ronda, this is Matt Wanty with Subroot. I'm following up on an email I sent about your team's phone conversations. It might be sitting in the big pile of other emails you're getting, I completely understand. I'll resend it Ronda and see if this is something you might find helpful. Have a great night!

Voicemails can be effective because the decision maker gets a chance to hear your voice. It gives them an opportunity to listen and further analyze if you're someone they want to speak with. If you're able to nail the voicemail, it gives you a chance to impress the decision maker and move past the first layer of rejection. They now know they're dealing with a professional sales person and you're not going to waste their time.

Email Tracking

The use of an email tracker can be very helpful when you're sending cold emails. If you didn't know, email tracking data lets you see when the receiver opens the email and how many times they opened it. Many sales reps monitor the tracking information so they can call prospects shortly after they've opened their email. Depending on the situation, this can significantly increase pick up rates. Of course, some decision makers may feel this is a little "big brother" but that can usually be overcome with a meaningful conversation.

In the cases where a decision maker opens your email a few times, you can try to use that to spark a response. Some reps in this situation like to send an email simply asking for feedback on that previous email. This isn't a bad strategy but I can tell you it doesn't work on me. When I get feedback requests I know that the rep's objective isn't feedback but really for me to engage. Other people may feel the same way. In this situation, I use an approach that's a little different. Like this:

Hi Rana,

I want to make sure I didn't confuse things in my last email. If there's anything I could explain better about Subroot, please let me know.

Best,
Matt

Personalization

Personalizing emails is a tough challenge for a lot of sales people. Since most reps are being held to lofty KPIs the time needed to personalize often isn't there. It's very time consuming, so some companies are starting to manage the process altogether. There is automation out there to help sales people personalize but I've never seen any rave reviews. I think personalization is hard enough to pull off when you're not trying to fake it with AI.

Another reason it can be difficult to personalize cold emails is because some decision makers don't have a web presence. In some cases, it's literally impossible to find any information on the decision maker. If you run into this, another option is to make the personalization about the company. You can research recent news about the prospect and try to connect it to the decision maker. Think about how the event could have impacted this person and then make that the personalization. For example:

Hi Ramesh, I can imagine you're swamped with all of your company's recent acquisitions.

Remember, the reason you want to personalize emails is so you can standout and break through that first layer of rejection. There are multiple ways to do it and with every cold email you want to use the one that gives you the best chance. If the personalization option isn't your best play, choose something else.

Another way you can standout with your cold emails is with statements that lighten the mood. Endings like "I hope this wasn't too painful" or "Either way, here's pic of a bunny to make you smile" can be attention grabbing. Of course, certain decision maker personas may not appreciate these types of cutesy approaches. (Most CFOs probably couldn't care less if there's a panda inserted in your email.)

Sometimes when I'm researching a prospect I'll make an unplanned decision to try and be funny in my outreach. Maybe it's their photo or something I read in their LinkedIn profile. I think if you're able to lighten the vibe of your email, it can go a long way with opening the door. Just follow your gut when it comes to things like this.

There are several gimmicks out there when it comes to cold emails. Make sure to avoid any methods that could be perceived as misleading or deceptive by the decision maker. Pass on things like positioning yourself as a buyer in order to create a conversation or other shady tactics like claiming to have met or spoken with the decision maker in the past when you haven't. Another questionable tactic to avoid is claiming that a prospect requested information from your company, when they haven't. These shenanigans might be effective in short spurts but they're never a long term answer. They will eventually hurt you and your company's reputation.

Subject Line & First Sentence

I'm sure you already know the subject line of your cold email is pretty important. It's a major chunk of the information the receiver will use to determine if they're going to open an email. In addition to the subject line, they may also see the first 10 to 15 words of your email. I think it's important to understand that most decision makers are looking at the subject line and the first sentence in an effort to eliminate an email. They'd prefer not to open it. How they perceive the subject line and first sentence is going to be the deciding factor. This is why subject lines that are generic or neutral can perform better than others.

Subject Lines

There are an endless number of ideas out there when it comes to cold email subject lines. Anytime I hear someone talking about a subject line that's working, I give it a shot. Why not? I don't usually get the same results but at least I'll gain more perspective on what's working for me and what's not.

One subject line that I use for my cold emails is "Question about (fill in blank)". The 'fill in the blank' can be their company name, a product they offer or anything else relative to them and my email. People usually don't mind questions and sometimes it piques their curiosity enough to open the email. Another way to keep it generic is by simply putting your company name in the subject line, *i.e.* Subroot. There is data out there showing that emails with a one word "company name" subject line can receive double the average opens.

Another popular approach reps are using right now is what I call the 'three subject' subject line. Like these:

Subject: Love using BigTech, programming ideas, cost reduction.

Or

Subject; Congrats on your growth, metric based recruiting, Evergreen.

This approach is almost the opposite of keeping it generic. Instead the subject line is very specific but by including three different subjects, it gives you triple the opportunity to break through the first layer of rejection.

If you've left a powerful voicemail; make sure your subject line refers to it. It will make it easier for the decision maker to connect the two sales touches. If they connect with your voicemail they'll probably read your email.

MailChimp, an email marketing service, states that the best subject line for open rates is the full name of the person receiving the email. For example:

Subject: Ron Swanson

Also, you may already know this but "Re:" emails get much higher open rates. If you've already emailed the prospect make sure to keep adding to the thread for at least the next few emails. Cold email is a number's game and you should always be optimizing to increase your open rates.

First Sentence

In the first sentence (or 10 to 15 words of the email), personalization can be really effective at helping you stand out. If the personalization connects with the decision maker you'll definitely get their attention. A common approach is congratulating the receiver or their company for something they recently accomplished. "Congratulations on your acquisition of BigTech." or "Congratulations on writing your second book." Keep in mind though, there's always a possibility other sales reps are using this same strategy. If the decision maker is getting this message often, it's going to put you in the same category as everyone else telling them, "Great job!"

Here are some other ways to get the receiver's attention with the first sentence. Humorous statements can go a long way in motivating them to open an email. Little quips can be risky but if you feel like you have something good, give it shot. This could be something like: ·

"Read any good emails lately?"

"Please don't hate me for emailing you?!"

"I used to think accounting was exciting."

Another thing you can do with the first sentence is to connect on a personal interest of the decision maker. This can be really effective at breaking through the first layer and getting them to check out your email. If you dig into their LinkedIn or other social media profiles there's usually something you can use. Things like:

"Thank you for your suggestion on reading Messy Middle, great book."

"I noticed you've been all over the world with Amazon, that's amazing!"

"Playing basketball overseas must have been an incredible experience for you, congrats."

By continually testing ideas you'll be able to get a good feel of what subject line and first sentence work best for you. If you're able to find something that consistently moves you past the first layer of rejection, you'll be in the driver's seat with cold email. Here are some other things to consider when it comes to the email's preview:

1. Always avoid braggadocios or buzzword subject lines like: "We're the No. 1 Choice", "Best Hybrid Technology", "Industry Leading Warranty", etc.

2. Adding words like "noticed", "read", "your" and "came across" in your first sentence, can help pique interest.

3. First sentences that find commonality can be effective, like: "John Sanders referred me" or "We share a common client in Nike".

Cold Email Complexity

Here are some complexities to consider when writing cold emails:

a. Most cold emails don't get read. Your email will need to stand out in order to entice decision makers to open it.

b. In many cases, decision makers are reading emails searching for information that's going to allow them to eliminate you. The more information you include, the higher the chance they'll find a reason to dismiss your email.

c. Getting higher "open" and "reply" rates doesn't necessarily mean an increased number of meetings or closed business.

d. The purpose of the email is to get them interested in your company but the email can't be all about you, you, you.

e. Links and attachments can impact security screening.

Getting a Response

Here are some reasons why decision makers respond to an email:

1. Timing - You've connected with something they're currently seeking.

2. Problem – You've exposed a problem they're having and would like to fix.

3. Referral – You've used a referral to effectively move past the first layer of rejection.

4. Improvement – You've gained their interest in getting better. (Things like more efficient, more sales, reduced costs, etc.)

Email Templates

In order to help structure your cold emails, this book provides some templates. These templates were created during an analysis of "high performing" cold emails published on the internet over the last ten years. We found several dozen emails so we broke them all down and eliminated the duplicates. This book includes four of the most common "high performing" emails that we identified, along with a live sample for each template. Remember, because these are lengthy emails, it's important that they're followed up with phone calls.

EVENT EMAIL

Event emails open with a specific purpose or event that prompted the outreach. Use this template to create your email.

Dear Decision Maker,

Personalized statement referring to an event or thing.

Question or statement that exposes a problem you solve.

Authority statement that provides a solution and establishes credibility.

Soft call to action.

EVENT EMAIL EXAMPLE: Development Company

Dear Ron,

Congratulations on your company's growth! I've noticed your development team is growing at an impressive rate.

You may be concerned that as you add IT resources your average cost of programming will rise. Have you ever considered outsourcing some of your development work?

Integrating with the right partner will reduce costs and increase flexibility. That's why BigTech has grown to over 300 developers and completed 1,250 projects in 40 different technologies.

If you're open to learning more please let me know.

Sincerely,

Sales Rep

PROBLEM EMAIL

Problem emails open with a specific problem that prompted the outreach. Use this template to create your email.

Dear Decision Maker,

Statement that exposes a possible problem.

Question or statement that magnifies the consequences.

Statement that provides a solution to the problem.

Authority statement that establishes credibility.

Soft call to action.

PROBLEM EMAIL EXAMPLE: Development Company

Dear Rana,

You may have noticed that measuring your development team's productivity can be a tricky business. As IT departments add resources, the average cost of development can sharply increase.

Of course, higher resource costs have a negative impact on your bottom line.

This can be avoided with the right development partner. Lower labor costs and increased flexibility provide a safer way to increase programming.

That's why BigTech has grown to over 300 developers and completed 1,250 projects in 40 different technologies.

If you're open to learning more, please let me know.

Sincerely,

Sales Rep

ASSUMPTION EMAIL

Assumption emails open by assuming the receiver is having a certain problem. Use this template to create your email.

Dear Decision Maker,

Statement that assumes the receiver is having a certain problem that your company can solve.

Authority statement that details a solution and establishes credibility.

Statement that explains the benefits of solving the problem.

Soft call to action.

ASSUMPTTION EMAIL EXAMPLE: Development Company

Dear Marley,

As your products continue to grow you may be experiencing an increase in your average cost of development.

This can be avoided with the right development partner. That's why BigTech has grown to over 300 developers and completed 1,250 projects in 40 different technologies.

Lower labor costs and increased flexibility gives development teams a more efficient way to grow.

If you're interested in learning more, please let me know.

Sincerely,

Sales Rep

REFERRAL EMAIL

Referral emails open by referencing a third party that gives you credibility with the decision maker. Use this template to create your email.

Dear Decision Maker,

Statement about referral and reason for the outreach.

Question or statement that exposes a problem that you solve.

Authority statement that provides a solution and establishes credibility.

Soft call to action.

REFERRAL EMAIL EXAMPLE: Development Company

Dear Eric Singer,

I was referred to you by Matt Wanty. He mentioned that you may be looking to reduce your development costs.

I'm sure you know that by integrating with the right overseas partner, you can significantly lower your average cost of programming.

That's why Big Tech has grown to over 500 developers and completed 2,500 projects in 50 different technologies.

If you're open to a short conversation about how we can help, please let me know.

Sincerely,

Sales Rep

LinkedIn Poll on maximum number of cold emails reps will send to unresponsive prospects – Sept 2020

Hey Sales People, what's typically the most emails you'll send to an unresponsive prospect?

1-2	10%
3-4	45%
5-6	26%
7 or more	19%

576 votes

Start Sending Cold Emails Like a SALES NINJA!

Eight of out ten decision makers prefer to be contacted via email.

General Email Open Rates

20.94 PERCENT
Hubspot

21.33 PERCENT
MailChimp

Email Campaign Open Rates

32 PERCENT
1 Email

46 PERCENT
2 Emails

50 PERCENT
3 to 4 Emails

43 PERCENT
5 to 8 Emails

Sending multiple emails to a prospect can significantly increase email open rates. The ideal number of emails to send is between 3 and 4.

Impact on personalizing email campaigns

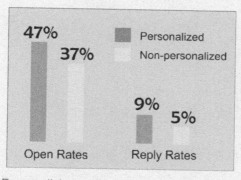

47%

37%

Personalized

Non-personalized

9%

5%

Open Rates

Reply Rates

Calling during a campaign can increase reply rates by over 200%.

Personalizing cold emails during a campaign can increase open and reply rates by twenty seven and eighty percent respectively.

Sources:
Hubspot - 28 Surprising Stats About Prospecting in 2020
Mailchimp - Email Marketing Benchmarks by Industry
Woodpecker - Cold Email Benchmarks by Campaign and Industry;
What makes Cold Emails Effective in 2020 (26,000 Campaigns)
Subroot - (1100 Email & Phone Campaigns)

 Subroot

9

LinkedIn Targeting

I think the most common way outbound sales people are using LinkedIn is to "send them a message and see what happens". This is the wrong approach and social media mindset. When it comes to LinkedIn and any other social media site, they're not just another channel of communication. Okay, I guess technically they're all communication but with social media the opportunity is far greater. The real prize is "familiarity".

Most sales people will never become an influencer; it takes a lot to achieve that. But that doesn't mean sales people can't use social media to become familiar to their prospects. The best way to think about this is what I call 'name cred'. Count how many times the decision maker may have seen your name. Once they've seen it eight to ten times, you may now be familiar to them. Any familiarity will significantly increase your chances of breaking through the first layer and getting your message across. Because at that point you're not just another sales rep. You're a name they know, someone they almost trust.

There are a few different ways to get in front of people on LinkedIn. Of course, your target decision maker's activity on the platform is what is going to determine if it can actually help. When you're using LinkedIn as a prospecting tool the first thing you want do with each target is to go to the section of their profile called "Activity". In this section, you're able to see if they post content or leave comments and Likes. Some decision makers fully participate on LinkedIn, while many others never login. Then, there are some decision makers who spend time on LinkedIn but never leave a trail. I call these people "wallflowers". You'd never know it but they're probably reading a lot of the same posts as everyone else. Sometimes, I think it's best to just assume your decision makers are there.

Connecting with Decision Makers

If your target decision maker shares content, then the easy option is to start commenting on their posts. Your comments should be supportive and add value. Definitely make sure to avoid any pitching. Since you're commenting on their posts, it's possible that the decision maker will end up sending you a connection request. If they don't, after a month of making comments and reposting some of their content, go ahead and send them a connection request. It's not necessary to include a note but if you have something meaningful to say it can be helpful. Remember, if there is any hint of pitch in your connection request, they're likely to decline it. They may even take the time to click, "I don't know this person".

If the decision maker doesn't post content but participates on LinkedIn with comments and likes, you should support their comments, while also providing your own insightful comments on that content. Don't be too obvious about it but remember your objective, 'name cred'. You're trying to get your name in front of the decision maker as many times as possible. Once they recognize you there's a good chance they'll accept your connection request.

What about the "wallflowers"? Since we're assuming they're active on LinkedIn and we just can't see them, it doesn't hurt to send them a connection request. If you have something good to offer, include a note but it's not always necessary. You can also send a follow up email to increase your chances of them accepting.

LinkedIn Poll on Connecting with Decision Makers on LinkedIn – Sept 2020

How often are you able to connect on LinkedIn with your target decision maker?

Never	4%
Rarely	37%
Half of the time	35%
Most of the time.	24%

303 votes

LinkedIn Messages

Once you're connected with a decision maker on LinkedIn, one of your options is to send them a direct message. Again, if you pitch in the first message it's probably going to turn them off. Most reps succeeding with LinkedIn messaging do a great job of connecting on a human level in the first interactions. These reps will avoid mentioning what they're selling or calls to action until they've established enough rapport with the decision maker. Like this:

Rep: *Hey Dave, I hope you had a great weekend. How have you been?*

Dave: *Good Matt. How are you?*

Rep: *Back to the grind, the weekend felt like it lasted 5 minutes!*

Dave: *I know what you mean, lol.*

Rep: *I should have known floating the river followed by a night on 6th street was too much for me, lol.*

As the dialogue continues the rep will eventually ask the decision maker if they're familiar with their company and provide some key value points. The casual dialogue can help push through the first layer of rejection making it more likely they'll consume the sales message.

Rep: *Hey Dave, are you familiar with Subroot? We help companies like BigTech get better results with outbound.*

Channel strategy, sales messaging, skills development are few of the areas we can help.

Other LinkedIn Options

Some sales reps use social visibility to try and catch the prospect with a warm call. When they see a target decision maker active on the platform, they simply call them. If the decision maker has seen the rep's name a few times the call won't feel as cold. This can increase the chances of a meaningful sales conversation.

LinkedIn Voicemail & Video

LinkedIn voicemail and video messages can both be very effective options when pursuing prospects on the platform. When compared to conventional voicemails, LinkedIn voicemail provides a much higher likelihood the decision maker will actually listen. Since you're engaging through LinkedIn you can use their profile to find details about them. Personal references are great things to mention at the beginning of your voicemail to increase the odds they'll continue listening. Again, you don't have to land your pitch in the first few interactions. Here's an example of a LI voicemail with a soft pitch.

Rep: *Good morning, Dave. I loved your recent post on KPIs! Thank you for that. It's refreshing to see a sales leader with that perspective. Anyway, I'm reaching out about your team's sales conversations; it would be great to give you a quick preview on how we help reps at companies like yours. Just let me know if you have the time to check it out. In the meantime, I'll keep enjoying your awesome content!*

LinkedIn video messages in particular can be highly effective. The view and response rates are the probably the highest you'll see in outbound sales right now. If the decision maker watches your video, any subsequent interactions will be warm. The fact that there's a high chance they'll watch your video gives you all the reason to make them. (I dig into LinkedIn video a lot more in chapter 13.)

LinkedIn Automation

If you're thinking about signing up for some type of LinkedIn automation, let me help with your decision. Don't do it. I get these automated messages on LinkedIn every day and you're wasting your money. This is a completely impersonal way to use a medium that is all about being personal. Some surprising results in this poll on LinkedIn mediums.

LinkedIn Poll on LinkedIn Mediums – Sept 2020

How are you using LinkedIn the most to reach decision makers?

LinkedIn Message/InMail	58%
LinkedIn Voicemail	2%
LinkedIn Video	10%
Sharing or engaging content	30%

221 votes

10

Direct Mail & Personal Gifts

When I think about all the sales people I've known doing well with outbound sales, I'd say about half of them use some form of direct mail or personal gift. Sending your prospect a physical item can be one of the most effective ways to push through the first layer of rejection. The mail itself can be as simple as a handwritten card or as elaborate as a personal gift. Sales people who are succeeding with direct mail are typically being either really creative or highly considerate. Sending a prospect something that is truly meaningful to them, it's a sure fire way to get their attention.

Promotional items and gifts can go a long way in breaking the ice but you also risk putting the decision maker in an awkward position. Some companies have strict policies against employees receiving gifts from vendors. If you run into this situation you can always just say something like:

"I'm so sorry, Ron. I didn't know. Since I have you, can I ask about…?"

Any direct mail or personal gift campaign is going to take an investment of money. Make sure to test the waters before you go all in a buy 1,000 bobble heads. But if you are able to come up with an idea that connects with your end buyer, it's possible for direct mail to perform 5 to 10 times better than other outbound sales channels.

LinkedIn Poll on Direct Mail Usage – Sept 2020

Hey sales people, choose the one that best fits your perspective on direct mail or gifts to open up doors?

I use it, works great!	23%
I use it, works so so	25%
I tried it, didn't work.	13%
I've never tried it.	39%

251 votes

Gifting Platforms

If you need help coming up with ideas, there are services like Sendoso and Alyce that are gifting platforms. They can help find and send the perfect personal gift to your prospects and customers.

Personal Gift Ideas

Personal gifts work great at breaking through the first layer of rejection. There are so many creative concepts out there I can't cover them all in this book. So here's some information that I think will be help you find the perfect gift for your prospects.

As you already know, if you want to send a complete stranger a personal gift you're going to have to gather some information on them. LinkedIn and other social media platforms are a great source for this information. Here are some things you can look for to turn into great gift ideas.

Children

Decision makers who have young children (especially babies) will usually find gifts like toys, stuffed animals, children's books, and games highly thoughtful. Raising babies is tough stuff and new parents will usually value any help they can get.

Hobbies

If you're able to figure out a prospect's hobby, sending them a related gift is a great way to shoot through the first layer of rejection. Golf, fitness, running, gardening and cooking can all be identifiable hobbies. Gifts like custom printed golf balls, training swag, gardening tools and cooking utensils can go a long way.

Pets

Oh, do people love their pets! Dogs, cats, exotic fish, turtles, snakes and many other creatures can be the apple of someone's eye. Pets are a great play when it comes to personal gifts. If you sent me a cat collar or some treats right now, you would get my attention. When it comes to pet goods you'll find a ton of different ideas for any type of pet. A quick search on chewy.com or barkbox should have you well on your way to getting some pet lover's attention.

College or University

Who doesn't like paraphernalia from their alma mater? Probably drop outs. Of course, that's not the case for most decision makers. You can easily find their alma mater listed on their LinkedIn profile and then search Google and find the perfect gift.

Charities

If you find the decision maker is involved with a charity, a donation will certainly move you past the first layer. If you think this concept is too "pay for play". Here's a newsflash for you, sometimes that's how life works. Sales people didn't make the rules; we're just playing the game and you have as much right as anyone else.

11

Cold Calling

Cold calling is still a very effective outbound sales channel for those who have mastered it (regardless of the naysayers). Many organizations get terrible results with cold calling because their sales teams are using ineffective strategies. Other companies, who have acquired most of their customers with cold calling, are able to do so because they've developed strategies that make cold calling achievable for their team.

Sadly, cold calling isn't always the right channel. Every industry is different and it's hard to predict when it's going to work. I can tell you with certainty; cold calling is most effective when you're trying to inform decision makers about something they're not aware of. It's the least effective when you're trying to convince them that you're the greatest thing in the world and/or superior to your competition. If you happen to be introducing a new product or service to the market, cold calling could be your best weapon.

Sending reps onto the phones with no game plan is like sending them into the witches den with no magic. New sales people usually need direction and a good strategy to embrace cold calling as a regular activity. The better the strategy, the more likely cold calling will be sustainable.

It's tempting for companies to load up large call volumes in order to push up the number of conversations for reps. In some cases this can be effective but it all depends on what you're selling and the target market. The last thing you want to do is waste your rep's time and mental energy on meaningless conversations. It's much better to have 3 to 5 conversations per day that result in a sale then 15 conversations with people who don't buy what you sell. Successful cold calling starts with good targeting.

Dialogue

Sales people who are able to get decision makers talking during a cold call usually have the best results. The easiest way to get a person talking is by asking them good questions. Anybody who is picking up the phone should be prepared with two or three questions that open up a dialogue with a decision maker. The questions that you created in Chapter 5 should work really well on cold calls. They may not work every time but they'll consistently create dialogue.

When you're having a back and forth conversation during a cold call, the chances of conveying your company's message increase significantly. Once decision makers are engaged, they're much more likely to absorb the message you're sharing. Long value pitches usually fail because they overload the decision maker at a time when they're not ready to listen.

Cold Call Opening

One of the most important parts of a cold call is your opening statement. It gives you a brief opportunity to impress the decision maker and let them know this is going to be a pleasant experience. Your statement should contain at least who you are, where you're from, and why you're calling. I understand that there are some trendy strategies aimed at withholding your identity until later in the call. Keep in mind, most real decision makers won't share information until they know who you are.

Cold Call Opening Example

Hi Ron, thanks for picking up, this Matt Wanty from Subroot. I'm calling about your team's sales conversations and I wanted to introduce you to our program. Ron, I should probably ask, is the phone even part of your sales strategy at BigTech?

Cold Call Closing Example

Hey Ronda, it seems like we're connecting on a few different things here. Could we set up a time when I can give you more details on how Subroot can help you solve these problems?

Connecting with Decision Makers

Okay, I know what you're going to ask. How can I get decision makers to pick up my cold call? One of the big complaints surrounding cold calling are the connection rates.

Reps can make hundreds of dials and only talk to a few people. Unfortunately, this is one of the realities of cold calling. But there are ways to improve your connection rates. One is by calling decision makers at the right time. I've seen reps connect three to four times more often because they're calling decision makers at an optimum time.

LinkedIn Connections

Connection rates are higher when you're calling people in your LinkedIn network. If a decision maker connects with you on LinkedIn, it should automatically become a call task.

Optimum Time

When is the optimum time? It really depends on the target decision maker and their market. Many people, including myself, will tell you that the best time to reach prospects is near the end of the day, between 3:30pm and 5:30pm in the prospect's time zone. Decision makers will be finished with meetings and usually wrapping up the day in their office. They could even be killing some time; making it easier for them to engage in a conversation with a sales person. Other parts of the day when pick up rates may be higher are morning, before 9:30am and around lunch time.

In order to find the optimum time for your decision makers I would encourage you to test different parts of the day. If you're not connecting with decision makers, don't just keep making calls at that same time of day. On the next page is a LinkedIn poll that shows at what times of the day sales reps are making most of their cold calls.

LinkedIn Poll on Cold Calling Times – September 2020

Hey Cold Callers, what time of day do you make a majority of your cold calls?

Before 10am	29%
10:00am - 12:59pm	33%
1:00pm - 3:59pm	27%
After 4:00pm	11%

748 votes

Local Numbers

There are several sales technologies available that can make it appear as if you're calling from a local number. This can increase connection rates because people are more apt to answer local numbers. Some critics view this as deceptive and you'll have to make your own conclusion. Most of the people I know who use local numbers seem to really value it.

Lure Email

There are other tactics that can help you get decision makers to pick up your call. The most effective strategy I've used to get people to pick up my cold call is a tactic I call the "Lure email". I happened onto this technique in 2016 when I connected with over a dozen high level decision makers at major corporations. Some of the connections I remember include: Sr. VP of Digital Marketing at United Healthcare, VP of Marketing at Travelers Insurance, VP of Digital

Marketing at American Express. This is a tactic I teach in the Smart Outbound program and reps have success with it.

 Shelby Whelihan · 12:49 PM

Hi Matt, I used your advice of emailing in the morning and telling them I was going to call yesterday and I got a live contact with someone I have been trying for months! Thank you for that nugget!

When I discovered the Lure email I was working for a technology company. We were testing out different things in order to get decision makers to pick up cold calls and the Lure email quickly became my go to strategy. This tactic has worked so well for me that I'm never surprised when the decision maker picks up.

Here's How It Works:

The lure email can be a mini-sequence within a bigger cadence. It consists of one email and one phone call. I've been successful with this strategy on every day of the week but you may find that certain days work better for you.

This sequence starts out with a very short email to lure the decision maker onto the phone. The email should reference one of your company's clients. Obviously, you'll need some clients in order to make this strategy work; the bigger the better.

Here are some examples when I was selling a technology that produced multi-lingual websites for major corporations. The company has several big name clients.

Lure Examples

Sr. VP of Digital at United Healthcare

Hi Dave,

I'm going to give you a call later today to explain how we help Humana manage their multi-lingual websites.

Hope we can connect,

Matt Wanty

VP of Marketing at Travelers Insurance

Hi Rita,

I'll be calling later today to explain how we help State Farm easily manage their Spanish website.

Look forward to connecting.

Matt Wanty

VP of Digital Marketing at American Express

Hi Rich,

I'll be calling later today to explain how we helped Fidelity increase traffic on their international websites.

I look forward to talking.

Matt Wanty

Lure Email Summary

Each of these Lure emails worked to get the decision maker to pick up my call. Even better, it didn't even feel like a cold call. Since they were expecting me, they didn't answer with a super gruff, "this is Rick!" They don't feel as interrupted because of the email. Most of the conversations I've had using Lure are ultra-productive because the prospects usually pick up ready to learn. Of course, this isn't going to work every time. The decision makers could be on vacation or in a meeting at the time of your call. Or they could just hate the approach. But the Lure Email has worked so well for me over the last few years that it will always be part of my outbound strategy.

This is a common question about Lure, "what if you don't have a current customer that's a direct competitor?" No problem. The next best thing is a customer that could be familiar to the decision maker. A client that's based in their same city or state will usually work well. Even a well-known company that tops your logo list can help lure them onto the phone.

One last note, the Lure email only works if it's a same day email and phone call. Sometimes reps like to turn it into "I'll call you later in the week". I'd only use the Lure email if you're going to do it all one day. Wrapping it all up in the same day, that's what makes this tactic work.

Cold Calling Tips

Here's a comprehensive list of tips to help you conquer cold calling.

1. Internalize your opening statement – The first few sentences that you speak on a cold call are the most important and the hardest to deliver. Practice your opening statement over and over so you're able to convey it under extreme duress. Because sometimes that's how you're going to feel.

2. Call when they answer – It sounds simple but most reps don't follow this advice. Figure out when your target persona is most likely to answer their phone and focus your calls around that time. Watch your connection rate significantly increase.

3. Master your value proposition – Everybody already knows this but few take it to heart. Whatever your prospects generally find valuable about your company's offering, that's the most important information you can share on a cold call. If you don't understand this information thoroughly it's difficult to be successful on the phone.

4. Closed-ended questions – Prepare two or three closed-ended questions related to the problem you solve. Closed-ended questions allow decision makers to warm up to the conversation. They also give the rep the opportunity to have a response prepared for either answer.

5. Be ready on every dial – Since a low percentage of our cold calls get answered it can be a challenge to stay mentally ready. You can keep yourself focused by going into every dial believing the decision maker is going to answer their phone. Literally tell yourself, "this call is the one".

6. Education trumps objections – A great way to move a prospect past an objection is to respond with a question that leads to education. For example, "We're not looking for new sales tech" said the decision maker. "I understand; do you know what our technology does?' asks the sales person.

7. Use dialing automation – With all the technology available punching numbers is a waste of time. Improve productivity by using some type of dialing automation (there are free dialer apps available).

8. Company name on caller ID – It's important to leave some type of trail that you called. If your company name keeps popping up on the decision maker's Caller ID, that's something they'll probably see. It might pique their curiosity just enough to pick up your call one day.

9. Pay for accurate contact information – Having good phone numbers for your targets will result in talking to more people. Instead of searching yourself, it's far more productive to spend money on a service that will give you reliable data. (Message me if you need a referral.)

10. Use emails to get decision makers to pick up – Timely emails can be effective at enticing decision makers onto the phone, especially if you're able to mention one of their competitors. Keep in mind that most decision makers scan emails looking for information to reject you, the shorter the better.

11. Call on the right level – In most sales situations, calling on the right level decision maker is paramount. But again many reps don't follow this advice.

12. Call in one hour blocks – Calling in one hour windows can increase productivity and make for a more enjoyable day. Most reps can complete 20 targeted cold calls in one hour, unless of course, there are a few long conversations.

13. Stand while you make cold calls (if possible) – Cold calling is a high energy activity and standing can improve concentration and results. It's also a great way to prevent a sore bum from sitting all day.

14. Have alternative objectives – Many reps go into a cold call with their focus on setting a meeting. While most decision makers will never schedule a meeting with a sales person. This clash of titans often results in a conversation where virtually no information is shared. On every connection, try to convey key information because you may never get another chance.

15. Don't give up – There is no rule in sales or cold calling that says you have to quit after two weeks. As long as you're not bombarding the contact, be persistent and it might just pay off for you. My

biggest deal was opened after 6 months of periodically cold calling the decision maker.

16. Create a 'must reach' list – There are many deals out there that will only be opened with a cold call. Keep a list of 'must reach' decision makers so you can make calls to them periodically. This will also help you keep up with your KPIs.

17. Cold callers only – If your sales team is sitting around non-sales personnel making cold calls, then that could be a problem. Sales people don't watch them write code or fill in spreadsheets. They shouldn't have to feel like people outside of the sales team are listening to them do their jobs.

18. Don't call too often – Treat cold calling a lot like dating, you don't want to scare off your target. If you call too frequently they'll be more reluctant to answer. Unless of course they pick up to tell you to 'buzz off'.

19. Avoid gatekeepers – The job of a gatekeeper is to keep sales people out so it makes sense to try and avoid them altogether. If there is no other way your

best option is to simply be human and ask them for help.

20. Mock Cold Call – Mock cold calling is a great way to get your feet wet on the phone. Live practice with realistic objections will have you as prepared as possible to have real conversations.

21. Celebrate good cold calls – Successful cold calls can be few and far between. If you're not celebrating them then you're missing half the fun of cold calling. After a successful cold call it should appear as if you scored a touchdown in the Superbowl, nothing less.

22. Let bad cold calls go – Every sales rep is going to encounter terrible cold calls because people are unpredictable (including us). The only thing you can do is learn from it and immediately let it go. Next dial.

23. No more than two voicemails – Most voicemails are deleted without one word being heard. Having to erase the same voice more than 3 three times is very memorable for decision makers, but not in a good way. Voicemails can help you draw people onto the phone, otherwise they're pretty useless.

24. Scan the decision maker's LinkedIn profile – The more information you have on your target the higher chances you have to succeed on a cold call. Gather information and make your calls intelligently because you owe that to the people you're interrupting.

25. Match the decision maker's energy – Some decision makers will pick up the phone like they wish you weren't alive anymore. Match that intensity with confidence as you let them know why you're calling. Always remember, your time is important too.

26. Don't take days off – Cold calling is a lot like shooting free throws or playing video games. A few days off and the rust could cause you to miss a big opportunity on the phone. Keep the skill sharp by making dials every day. This will also help prevent call reluctance.

27. Learn with every conversation – The best way to learn is to ask questions. No matter who picks up your call, ask them a few questions and get more information about your prospect. Everything in sales is just a guess until you speak to somebody.

28. Don't sell the meeting – When you sell the meeting you share no information about what your company does. If the decision maker resists (most do), they leave without knowing any of the reasons why they should do business. If you happen to work for a company that sets meetings for other companies then I get it, you should sell the meeting.

29. Don't make cold calls all day – It's not healthy or sustainable. Make targeted cold calls in one hour blocks, 1 to 3 times a day. Consider No. 2 and schedule your cold calling blocks and prospecting ahead of time. Always know what you're doing tomorrow.

30. Don't hang up without a next step – Most conversations don't end in a meeting but that doesn't mean you can't keep the ball rolling. If the prospect isn't disqualified, stay in front of them with an occasional email or phone call. Some deals take years to develop and since you've opened the door with your cold call, keep it open.

31. Always send at least one email – Since some decision makers will never pick up your cold call,

it's important to send at least one email to check your timing. Personally, I like to make a couple dials into the prospect before I give them the opportunity to reject my email.

32. Spend time creating warm calls – Cold calls are great but with LinkedIn it's possible to create warm calls. If your targets are active, read their comments and everything they post. Try to be supportive but not obnoxious about it. After a handful of social touches try giving them a call.

33. When you're in a groove making cold calls they're easier to face. That's why I encourage people to make dials every day. Here's what I do when I'm struggling to get back on the phone. Anytime I have an important call scheduled with a prospect, I block out an hour or two afterwards. Right after the meeting is over, I start making cold calls. The tension isn't as high and your words flow better.

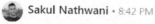
Sakul Nathwani · 8:42 PM

Hey Matt, Wanted to let you know that this post from a couple months ago made a big difference and allowed me to close another deal. I had a prospect who I was hesitant to call for a few different reasons. But after closing a different deal, I used my momentum to call him and a couple weeks later the deal was closed.

Here's a LinkedIn poll on the average number of dials sales people are required to make each day.

LinkedIn Poll on Cold Calling KPIs – September 2020

Hey Cold Callers, how many dials are you required to make each day?

None	13%
1-50	47%
51-100	29%
100+	10%

775 votes

12

Demand Generation Marketing

Why the heck is there a chapter on Demand Generation Marketing in a book about outbound sales? Because it's important and you need to know about it would be my answer. As an author and outbound sales consultant, Demand Generation Marketing is how I get most of my clients. I know, you're not selling a book or consulting services. You should be happy about that. ☺ What's important about Demand Generation Marketing to outbound sales people is that in actuality, it's what we do. We create demand that isn't there. We do it with cold calls, cold emails, messages, direct mail, video messages, sales conversations, referrals, etc. In marketing, the demand generation is done over time with advertisements, social media posts, webinars, podcasts, eBooks, guides, white papers, etc. We share the same goal of demand generation; we just use different modes to achieve it.

Having covered that, come along with me on a journey and relax your thoughts for a second. Think about decision makers and what it would be like living in their world. They're busy. They're accomplished. They're bombarded. They're content. Things are usually working just fine;

they've worked really hard to make sure of it. They're not seeking solutions because they don't believe they have any problems. Enter demand generation marketing. With demand gen marketing, the long term objective is to show these decision makers that they do have a problem or that they could be doing things better. But it's not just the education that's important. It's the credibility that you gain when you're the one teaching them. That's the power of demand generation marketing. That's the reason they're going to come to you.

Now, as a salesperson should you write an eBook? Probably not. But that doesn't mean you can't use demand generation marketing as part of your process. Instead of an eBook or webinar, your continuous gift as a salesperson could be providing intelligence. Can you supply your target decision maker with information they value? Like market data, industry news or something they'd find interesting. The best information is going to help show that there could be room for improvement. Here's an example:

Hey Ronda, did you know thousands of people in Europe are searching Google for your product? Here's some recent data that I thought you might find interesting.

Here's the bottom line, if you're able to provide prospects with a series of helpful information you should definitely be doing it. Because you're helping them, this approach makes it really hard for decision makers to reject you at the first layer.

Start thinking about the type of intelligence you can share with your prospect. Here are some examples. Realtors use demand generation marketing by providing information to their prospects that will help them raise their property values. Insurance companies provide information that helps their prospects make their homes and businesses safer. Accountants give away free tax advice. Health clubs provide prospective members with free information on how to maintain a healthy lifestyle. Sales trainers provide free tips on selling. Financial advisers share information to help their prospects manage their finances. Lawyers, I don't know what lawyers do actually.

Social media provides the perfect platform to execute demand generation marketing as a salesperson. There are many sales people out there right now doing it with written posts, document shares, videos and comments. I'm not saying you should spend most of your time creating content because that's probably not the case. But if you're interested in sharing, then you should at least give it a shot, even if it's in your free time. The most common mistake I see sales people make is producing content for other sales people when their target decision maker is in IT or accounting. If you're going to give it a shot make sure the content you're sharing is something your target decision makers will find valuable. Otherwise you're just building a personal brand, which isn't a bad thing, but it's not going to help you hit your quota.

Social Media Writing Tips

One of the best ways to be seen is by writing popular comments, posts and articles. Here's a list of tips that will help if you want to try generating leads by writing on social media:

1. Read. If you want to write to the public then you should be reading. You can read posts, articles, books, any of that stuff will help you become a better writer. But if you want to experience writing that will take your skills to the next level, read the daily news. Reporters take any topic and turn it into clear, concise information that everyone can understand. That's a powerful writing skill to have on social media.

2. Write. If you want to write to the public then you need to be writing. Not only what you post but what you would never post. Writing is like any other skill, if you invest time and effort you can improve your game. You don't have to be a grammar master or a spelling bee champ either. Keep in mind, the most appreciated writing on social media is usually very easy to read.

3. Choose an audience. The point of sales people writing on social media is to be seen and heard by their target market. In order for this to be productive, you'll need to attract this audience by adding value in topics they find important. Establishing yourself as

an expert can make it really easy for people to buy from you.

4. Make sure it's helpful. Your litmus test for every post on social media should be, does it help? The entire world doesn't need to love it but if someone in your audience finds it useful, then it works. If you realize you're in the middle of writing something that doesn't help, just keep tweaking it until it does.

5. It can't be about you. If everything you write is going to be a semi- autobiography then you should just stop reading these tips. People hate self-centered, keep it about them and splash your ego in here and there. Stay true to helping people and you can't go wrong.

6. Get a proofreader. Preferably someone who is proficient in grammar. Sadly, everything you release to this world is a reflection of your intelligence. Bad grammar, misspelled words; the haters eat that up. Don't give them the chance by finding a solid proofreader. Work with them to perfect every post or article because you owe that to your readers.

7. Rarely sell. Since the road to success on social media involves being helpful, writing about your product or service can send the wrong message. It's definitely okay to pitch once in a while but always err on the side of less. You'll probably find engagement drop in proportion to the size of your pitch.

8. Make well written comments. Success on social media is a two-way street of content creation and support. Find other content creators in your space who you can support because you AGREE with their views. Engage with their content by writing quality comments that add value, they'll usually return the favor. You'll know you're headed in the right direction on social media when people start supporting your comments.

9. Consistency. If you want consistent results you'll have to be a consistent writer. An hour or two a day will usually produce plenty of content for social media. But you have to stick with it through the good times and the bad hangovers, your audience will require that.

10. Write about real life. The best content usually comes from people's daily life. You might find what you're doing every day to be one of the easiest things to write about. If you're making cold calls, meeting prospects, working trade shows, reading a book, testing new products, traveling, etc., ideas will probably be popping in your head. Start writing them down and you'll have a continuous stream of new content.

11. Listen to comments. If you're successful writing on social media there are a few people who will hate

you for it. Make sure to completely ignore them but listen to other people's comments. Positive and challenging voices will help by giving you feedback, providing more topics for you to write about and just supporting you in general. People will definitely let you know if your writing is helpful.

12. Be real. Keep your writing real and your audience will love you for it. The best content is usually based on real situations to which people can relate. Real failure and emotion does really well on social media but there's a fine line to coming off needy or fake. One of the worst things you can do is misrepresent yourself or your accomplishments.

13. Create a style. Every writer has their own style, even non-writer sales writers. Don't be afraid to find a different way to write your content. As long as people can understand what you're trying to say, they'll appreciate the creativity.

14. Shorter is always better. We're sales people and our message should always be concise. If your writing tends to drag on, have a lot of unnecessary words, or repeat information, people probably won't like that.

Everybody's impatient, cleanly written content is appreciated and rewarded. During the proofing stage you should be trying to remove words.

15. Read it out loud. Reading your writing out loud (at least in your head) is a great way to test its readability. If it doesn't flow for you then it won't be easy for other people to read either. One of the harsh realities about writing is that if people find it difficult to read, they'll quickly move on.

16. Research facts. You can't afford to spend time doing research for everything you write. However, if you decide to quote factual information you should make sure it's accurate. A quick Google search will probably help confirm with multiple sources. There's a good chance the trolls will be investigating before they're even finished reading.

17. Be humble. Elitism is an ineffective position on social media. Writing in a manner that makes you out to be above others probably won't get you very far. It will be much easier to capture an audience if you're relatable and adding value to their lives.

18. Avoid bad language. There are many people successful on social media who use explicit language but they're probably not sales people just starting out. Keep your writing clean for the most part and you'll increase your odds of success. Using profanity would probably offend more decision makers than it would impress.

19. Pay attention to algorithms. You're not going to crack the algorithm but at least pay attention to what's happening around you. If your posts seem to work better on a certain day and time, use that to your advantage. Always take a look at what's going on in the feed before posting your content. If engagement seems low, it may be better to hold your post for a busier time.

20. Tee up your best content. If you feel like certain things you've written are better than the rest, post them on the days with the most engagement. Since it's important to be consistent, use your weaker content for times when traffic isn't as heavy. Try to never let writing that you're proud of get lost because of poor timing.

21. Don't get full of yourself. If you're successful writing on social media the payoff will be new business opportunities. One of the quickest ways to negate any progress is to let it go to your head. If you find yourself writing about your successful writing, then it's probably gone to your head.

13

Video Messaging

Lights, camera, action... Salespeople are using video messaging as a prospecting tool, watch out world! Why is video more effective than email at breaking through the first layer of rejection? The first thing you should know is that the human brain processes visuals much faster than it does text. Around ninety percent of the information we transmit is visual. In addition, our brains can process visuals at a rate 60,000 times faster than it does text.

This means the recipient of your outreach will consume a lot more information by watching a video than they would by reading an email. If that wasn't enough, with a video the decision maker gets to see your body language, which can make up 55% of conversational communication. With all that being said, video messaging has significantly enhanced a salesperson's ability to convey their message.

When it comes to using video messaging, like everything else in outbound sales, it starts with getting the prospect to watch the video. Many reps complain that very few of their videos are actually watched by decision makers. I've run into reps with view rates lower than 5%. The email's subject line, first sentence and the video's thumbnail are the main factors determining whether a decision maker will watch. The bottom line, in order to make video messaging effective you have to figure out ways to get decision makers to watch it. If you can find something that consistently works then you have a great chance at getting results with this medium.

Video Content

The content of the video shouldn't be much different than what you would write in a cold email. Since the first few seconds of the video will likely determine if they watch the rest, personalizing your introduction can be helpful. The video should have a strong purpose, include a statement or question that exposes a problem and provide value information on how your company can make things better. Most video messages I've seen are around one minute. I think that's a great length and I'd try to keep it under 90 seconds if possible.

Personality

Another great thing about video is the decision maker gets a chance to see your personality. If they like you then it will instantly push you through the first layer of rejection. One of the most important things to remember when making videos is to be you. Mistakes and other imperfections can actually work in your favor. Here's a sample framework for a prospecting video but don't be afraid to test out other ideas.

Video Framework

First 10 seconds: Intro & Personalization

Next 10 seconds: Purpose of your video

Next 15 seconds: Question/Statement exposing a problem

Next 15 seconds: Value statement explaining how you help

Last 10 seconds: Soft close

Video Tools

Now, you can't just shoot a video and attach it to your email. The file size is way too big. The best chance you have to get someone to watch your video is if it's embedded right in the email. There are several freemiums that provide this service like Vidyard and Loom. These tools also make it easy to create videos with a screen capture, allowing you to give decision makers a visual presentation. This can be especially effective because it allows you to present more detailed information. Since it's a video the decision maker can rewind if they'd like to see something again.

LinkedIn Video Messages

Sending video messages through Social Media is a completely different thing. As you probably know, you have to be connected with the prospect, which can be a major stumbling block with some personas. If you're able to connect, LinkedIn video messaging is red hot right now. LinkedIn doesn't let you know if someone watched your video but view percentages must be really high because response rates can as high as 90%. But again, you need to be connected with the decision maker which will limit the number of prospects you can reach with this option. Here's a common mini-cadence for LinkedIn video prospecting.

LinkedIn Video Cadence

Day 1: Connection Request

Day 2: Personalized Email

Day 5: LI video message

Video Production

As far as making the video, do you have a smartphone? That's all you need to make it happen, a bunch of fancy equipment is not necessary. Creating these videos can be a little difficult at first but once you get rolling they should become easier and easier.

If you're looking for an eighteen step process on making video messages for prospects then you're probably overthinking it. The best productions seem to be short, natural, personal and candid. Of course, don't wake up with bedhead and shoot a video but you don't want to overthink the background either. If you're making videos to share on social media then you might want to set up a nice background. But for sales related video messages, I think the best ones are when the rep takes up most of the screen sitting or standing with a plain background. Here are two great examples.

Louis Weimann

Anjali Purkayastha

14

Random Tactics

Here are some random tactics that keep showing up on my radar. These are effective ways to move past the first layer of rejection but always keep your buyer persona in mind. Some of these tactics may fit with your target decision makers and others probably won't.

Custom Graphic Email

If you want to see your cold email open rates increase handsomely, start including a custom graphic with your email. Graphics are easier for the recipient to view than text is to read. Of course the graphic needs to be pertinent and interesting but it also should be customized for each prospect. If you're able to capture their attention with a graphic, then you'll move past the first layer and they'll read your email. This tactic is most effective if the email and the graphic work together to convey some of your company's key value information.

One of the best ways to do this is with a screen shot of your product or service in action. If you can reference a current client at the same time it will give you credibility as the prospect visualizes exactly what you're selling. The graphic could be a screen shot of a pertinent website, software in action or valuable charts on the industry. Other options could be infographics and productivity tools useful to your target decision maker.

Tag Team Prospecting

The technique of team prospecting has been utilized for decades. Whether it's a sales rep and manager, or an SDR (Sales Development Rep) and an AE (Account Executive), two voices knocking on a prospect's door equals more attention. Decision makers can be more compelled to meet once they hear from a senior representative.

The easiest way to implement this strategy is for the rep to CC the manager or AE on any emails. Then they can chime in with more info on what they think will interest the decision maker. If your company uses a sophisticated phone system then the rep should be able to bring the manager or AE right onto a call. On the flip side of this, having two sales reps unknowingly calling on the same company can be a confusing turnoff for prospects. Don't let that be you.

Face Time

I've come across some reps having good conversations by face timing decision makers. If you have an iPhone I'm sure you know that you can tell if someone else does as well. If the decision maker's cell is an iPhone then you can try face timing them. The tactic isn't always met kindly but when it does work it not only opens the door but it can also advance the sales cycle.

Memes

A similar tactic to the custom graphic email is a relevant meme. If you can come up with a creative idea and really nail the short message, memes can be a great way to move past the first layer of rejection. There are an endless number of ideas out there when it comes to memes. If you're interested in pursuing this tactic I'd search Memes and try to come up with some relative ideas. Pass the ideas by your friends and colleagues first, before you let them loose. If you find one that's really working you should definitely let me know.

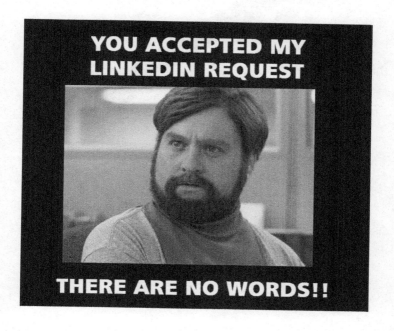

Personalized Graphics

If you want to try something that's basically free and can make a huge impact, you should consider personalized graphics. This can be as simple as a photo of a note like the one I received from Mark McInnes. Mark made a huge impression with this graphic and I'm sure you could think of a few ways to get noticed simply by taking a photo and sending it.

Breakup Emails

The title of this section probably sums up how I feel about breakup emails. To this day, I've never heard of anyone being successful with one. Giving someone who doesn't care about you an ultimatum doesn't seem like the best strategy. Instead of breaking up with a stranger who hasn't acknowledged you exist, send them an email telling them you're going to be "giving them a break".

Hi Ron,

It doesn't seem like this is the right time to discuss your team's sales conversations.

I'll give you a break and try to connect with you later in the year.

Thanks,
Matt Wanty

Caricatures

I've had a few reps tell me they were successfully opening up doors by sending decision makers a caricature of their LinkedIn pic. They would commission an overseas vendor to affordably create the caricature, between $5 and $15. Then they would simply send the decision maker a short note with the attached caricature. This type of gesture will usually shoot you through the first layer of rejection.

Cartoons

Another great way to capture the decision maker's attention is with a custom cartoon. They might be a little more expensive than the caricatures but you can convey a lot more with a cartoon. Check out this example from Cartoonist Richard Russell, he created it for one of my LinkedIn posts on cold calling.

By Richard Russell (cartoonrich.com)

Here's the post:

Calling all the brave sales people still making cold calls while...

sitting in the corner of their bedroom hoping the dog doesn't bark.

Or outside on the porch so they can't hear the baby crying.

Or sweating in the attic away from the neighbor's crappy speakers.

Making cold calls isn't easy.

What you have to go through right now is a whole other level of difficult.

I applaud you.

Your companies applaud you.

Keep up the great work!

#coldcalling #hard #covid

15

Decision Makers, Gatekeepers & Targeting

Since this is a book about outbound sales it probably makes sense to talk about the people we're calling on. One of the biggest differences with successful sales people is that they spend most of their time targeting and talking to the right people.

Sometimes those right decision makers are at the C-level and sometimes a Manager can make things happen. I've seen enough scenarios to know that it makes sense to call on all levels of an organization. But hold on, I'm not talking about carpet bombing everyone at the company. That's the last thing you want to do. I'm talking about a strategic approach based on profiling the prospect's hierarchy. By using LinkedIn, it's easy to gain a clear perspective of which people may be involved in a certain decision. This research takes some time but in the end it's well worth it. The first step is to find a person at the company who you think could be the primary decision maker. On their LinkedIn profile, go to the far right side to the section called "People Also Viewed". Check the section to see if there is anyone else from the company who you think could also be involved. Keep repeating this process with each contact until you have a good picture of the decision making hierarchy.

By systematically approaching a company you can make sure you're not getting a "No" from someone who isn't authorized to give it. What does systematic mean? First, figure out who you think are the top two or three decision makers. (In some cases you may decide it's just one person.) Then target these people first. I understand this may not help with your KPIs though, so you need to do what you need to do to maintain those.

If you don't make contact with one of the top decision makers you've identified, expand your outreach to other levels of the company. If you make contact with someone who is not a decision maker, quickly change the objective to gathering information. Here's a quick breakdown of what you need to know about the different levels within companies. Keep in mind, these are generalizations and you may run into situations that aren't anything like this.

C-Suite

The C-level can be tough to crack but if you can get your foot in the door, you could be sitting pretty. I've called on many C-level folks over my career and usually one of two things happens. Someone else tells me to buzz off or it's seems like they're a ghost. In many cases the people in the C-Suite are just like any other decision maker. If you can solve a problem or help with something that's important to them, you'll have their ear. I've found a great way to reach this level is with a Saturday morning email.

Sometimes if a Vice President ignores my outreach, I'll put my focus on the C-level. *"Hey CFO, I couldn't get ahold of Rick and I wanted to make sure you guys knew about a cost saving opportunity..."* or *"Hey COO, I was unable to connect with Rita and I wanted make sure you guys were aware of this new service..."*

Vice Presidents

I believe Vice Presidents are the most productive level that you can target as a sales person. The Vice President's job is to make things happen and sometimes changing to whatever you're selling is that thing. VPs are usually looking for ways to get noticed, that's part of how they got to that level in the first place. I kid. VPs are smart people and if you're able to show them how you can help, most of the time they'll listen. A big win for any Vice President is bringing on a new vendor that helps their company improve. If you can connect those dots and align your messaging, you will have success calling on Vice Presidents. Most VPs are super busy so it's best to keep your messaging all business.

Directors

In my experience, I've seen a lot of Directors who were virtually powerless. They admittedly couldn't approve a box of chocolates. But I've also seen Directors who could make things happen. I usually put my initial focus on Directors if there isn't a Vice President for that specific department. In this case, a Director will usually have the VP's decision making capacity.

One way you can tell if a Director may not have much power, they'll be one of multiple directors reporting to the same Vice President. If there's a pool of Directors, then the Vice President is probably calling all the shots. They'll need to maintain supremacy over those eager Directors. Kidding, again!

My experience with Directors is a little different than this LinkedIn poll. In this poll almost forty percent of the participants said they focused their cold calling efforts on Directors. However, another forty percent of the reps are still calling on the C-Suite and Vice Presidents combined.

LinkedIn Poll on Cold Calling Decision Makers–September 2020

Hey Cold Callers, which level decision maker receives the majority of your cold calls?

C-suite (President/Owner)	21%
VP level	22%
Directors	37%
Managers	20%

566 votes

Managers

It depends on the organization, the size of the transaction (and several other factors) but sometimes Managers can get it done. Definitely not every time, but it happens enough times to take them seriously with your outreach efforts. Managers usually have one foot on the frontlines which can give them a lot of power. In some departments it's also possible that a manager is the highest level title. In this situation, I would start my outreach with them and go from there.

However, it's also very possible that a manager has very little or no decision making power, especially when it comes to spending money. I've spoken with a few dozen sales managers over the last year that weren't empowered to purchase anything. Either way, if you get a chance to communicate with managers they're a great source of information and they'll usually tell you if they have any say. Don't be afraid to target managers if you're not having any luck elsewhere.

Gatekeepers

Ask for help. Respect their role. Follow up. That's all you can do with gatekeepers. I hate to be short with it but I'm not going to give you some cockamamie trick that works with gatekeepers. They don't exist. Their job is literally to be a gate. Rarely are you going to get anywhere with a gatekeeper but as a respectable sales person you still have to try.

Remember, many gatekeepers are going to tell you what you want to hear. It just makes things easier. Even if they do get your message and are sold, don't expect them to be able to convey it to the decision maker. In some cases they're putting their neck out there by even trying. If you run into gatekeeper always try to make it worth your while by getting as much information as possible.

Here's something I like to use if the gatekeeper is trying to shut my outreach down.

Gatekeeper: *"John's not available. You should send him an email; it's really the only way to reach him."*

Me: *"Thank you, I understand. I actually sent him an email but it's difficult to convey what we do in a message. Would you know if there's any time of day when it's possible to catch him? This is something John might want to hear more about."*

Or

Me: *"Thank you, I understand. I can try sending him an email but that might get me put in the wrong category. Do you know if there's any time of day when it's possible to catch him? This is something John might want to learn more about."*

Targeting

You might have the greatest outbound strategy in the world. The messaging can be perfect. The mediums you're using could be highly effective. But if you're calling on the wrong prospects then you're completely wasting your time. One of the biggest revelations that companies and sales people can have is identifying their best prospects. Some people like to call it ICP, that's 'Ideal Customer Profile' and not 'Insane Clown Posse' (which I once thought).

It's common sense. Spend most of your time selling to the right companies and you'll have a much greater chance for success. But it's not that simple. Identifying good prospects can take training, sometimes a lot of it. This is not something that's always provided by companies. Plus, in some markets identifying an ideal prospect can be like finding a needle in a haystack. All these factors can make it difficult for sales people to fulfill their KPIs. This is why "less than ideal" prospects usually find their way into a rep's pipeline. Hopefully you're not in that type of situation. Here are some tips on targeting the right prospects:

1. Evaluate current customers. Your customer's competitors or similar companies may also fit into your ICP.

2. Ask the best reps at your company about specific prospects. *"Hey Dave, do you think this company is a good prospect? Why not?"* After a dozen of these interactions you'll have a good perspective.

3. Challenge the norm. Sometimes the ICP a company has identified for itself makes total sense. Sometimes it doesn't. If you feel strongly that other sectors could benefit from your product or service, challenge things a bit.

Sales Navigator

LinkedIn's Sales Navigator is a great tool for targeting. It allows you to see the data in a way that other providers can't match. Sales Nav works much like any other database search tool but because it's tapped into everyone's profile, it provides much more insight. I'm not a LinkedIn Sales Navigator expert but I think I get the most out of the tool. Here are some Navigator search options that I think every sales person should know.

With every Navigator search you can see who has changed jobs in the last 90 days, who has posted on LinkedIn or who has been mentioned in the news in the last 30 days. Plus, if you dig into your target's profile, you can check out if they're following your company or any of your customers. All of this information can be very helpful when you're trying to push through the first layer of rejection.

Intelligence Tools

There are many great tools available that can help you identify prospects and gather intelligence. Job boards like Indeed or Glassdoor will give you information on what positions companies are trying to fill. There are other sites, like BuiltWith that give you details on how a company's website is built. While sites like Owler provide you valuable information on company revenues, capital raised and most importantly, a list of competitors (your prospects). There are also tools like CheetahIQ that provide deep intelligence on contacts like press releases, job postings, 10k mentions, podcast appearances and more. Whatever you sell you should be making sure that you're tapped into all the different resources available around it. Don't just follow the norm. You'll separate yourself from the rest of the sales team if you can find better prospects.

16

Conclusion

This might get a little deep. I think it's important for anybody involved with outbound sales to understand that it's a really difficult job. It's filled with rejection after rejection from prospects, customers and at times even your own company. The sheer fact that you would be willing to work in outbound sales makes you a special person. In reality, it's a job that most people would walk away from. Without you, some companies and industries wouldn't even exist. As a front-line rep you have the challenging task of taking a prospect from "zero to interested". The importance of your role at a company cannot be overstated.

Now that we got the pep talk out of the way there's something else you should know. Nobody cares. Nobody cares about your service or product. They couldn't care less about your quota. They don't give a crap about your customers or how many awards your company has won. They care about themselves. As an outbound sales person, everything you do should keep this one fact in mind. Your sales messaging and strategy should be tailored around how much your prospects literally don't care about you.

Now, back to the whole 'outbound sales being a really hard job' thing. You probably already knew that or you wouldn't have purchased this book. But you might not know about something that far out-weighs how many hours you work or the number of prospects you dial. What matters the most when it comes to taking on the challenging job of outbound sales is that you CARE. As Kenny Madden, LinkedIn Legend says, "You need to give a shit!" (His words.) In order to succeed with outbound you have to care 14X more than anyone else. More than the operation's people. More than the accounting folks. Way more than marketing. It's the total belief in what you do and what you sell that will help you crush outbound sales.

Okay, it's that time when I should probably try to sell you something else in my Clickfunnel. But I'm going to skip that part because this book should give you the tools you need to succeed with outbound sales. By testing, executing and perfecting these methods you should get to where you want to go. If for any reason you think I could be of further assistance, please feel free to email me: matt@subroot.com (or contact me on LinkedIn). I hope you enjoyed the book.

Best of luck with outbound!

The End

About the Author

Matt Wanty started his sales career as one of the top reps at a major logistics corporation. In 6 ½ years he sold over $20 million in new business before moving on to start his own company. Matt's first success as an entrepreneur was driven by selling a seven figure contract to one of the largest manufacturers in the world, 3M Company. He currently provides outbound sales consulting through his new company, Subroot.

Matt has helped reps and sales leaders at companies like Zoom, Salesforce, Seismic and Northwestern Mutual have more meaningful sales interactions by refining their messaging.

Matt also helps companies and reps formulate outbound sales strategies and enhance selling skills through Subroot's "Smart Outbound" program. You can follow Matt's content at linkedin.com/in/wanty where he reaches millions of sales people around the world.

EXTRA WORKBOOKS

Optimum Purpose Workbook

Who or what benefits from what your company provides?

[ ]

What is the benefit?

[ ]

Optimum purpose:

> I'm contacting you about...
>
> _____
>
> _____

Creating Questions Workbook

What problems does your company solve?

1.

2.

3.

How does your company specifically solve these problems?

1.

2.

3.

If you have a highlighter handy, go ahead and highlight the main point from each problem or solution.

Now create your questions. Start the sentences with things like: *Are you having any problems...? Do you think your team...? Are you using...? Do you have...? Do you feel like...?*

1.

2.

3.

~ Notes~

~ Notes~

~ Notes~

~ Notes~

~ Notes~

Made in the USA
Monee, IL
16 July 2022